T0182071

Lecture Notes on Mathematical Modelling in the Life Sciences

The rapid pace and development of the research in mathematics, biology and medicine has opened a niche for a new type of publication - short, up-to-date, readable lecture notes covering the breadth of mathematical modelling, analysis and computation in the life-sciences, at a high level, in both printed and electronic versions. The volumes in this series are written in a style accessible to researchers, professionals and graduate students in the mathematical and biological sciences. They can serve as an introduction to recent and emerging subject areas and/or as an advanced teaching aid at colleges, institutes and universities. Besides monographs, we envision that this series will also provide an outlet for material less formally presented and more anticipatory of future needs, yet of immediate interest because of the novelty of its treatment of an application, or of the mathematics being developed in the context of exciting applications. It is important to note that the LMML focuses on books by one or more authors, not on edited volumes. The topics in LMML range from the molecular through the organismal to the population level, e.g. genes and proteins, evolution, cell biology, developmental biology, neuroscience, organ, tissue and whole body science, immunology and disease, bioengineering and biofluids, population biology and systems biology. Mathematical methods include dynamical systems, ergodic theory, partial differential equations, calculus of variations, numerical analysis and scientific computing, differential geometry, topology, optimal control, probability, stochastics, statistical mechanics, combinatorics, algebra, number theory, etc., which contribute to a deeper understanding of biomedical problems.

More information about this series at http://www.springer.com/series/10049

Takashi Suzuki

Mathematical Methods
for Cancer Evolution

 Springer

Takashi Suzuki
Center for Mathematical Modeling
    and Data Science
Osaka University
Toyonaka, Osaka
Japan

ISSN 2193-4789   ISSN 2193-4797   (electronic)
Lecture Notes on Mathematical Modelling in the Life Sciences
ISBN 978-981-10-3670-5   ISBN 978-981-10-3671-2   (eBook)
DOI 10.1007/978-981-10-3671-2

Library of Congress Control Number: 2017938548

Printed on acid-free paper

This Springer imprint is published by Springer Nature
The registered company is Springer Nature Singapore Pte Ltd.
The registered company address is: 152 Beach Road, #21-01/04 Gateway East, Singapore 189721, Singapore

Preface

Higher order living things consist of many key components such as the skeleton, the locomotor apparatus, the respiratory apparatus, the cardiovascular system, the digestive system, nerves, ..., and then the central nervous system, which control them all in unity. Organs, tissues, and cells form the organic hierarchy of an individual, while each of them exhibits a certain individuality. *Individuality* is thus a central factor of life.

Cells can be thought of the basic units of individuality. For example, cancer cells invade normal tissues to destroy organs, viruses invade normal cells, and immune cells attack viruses. Inside a cell are the nucleus, microsomes, Golgi apparatus, lysosomes, mitochondria, ... and various other organelles, through which proteins transmit signals.

The purpose of the present monograph is to describe some recent developments in mathematical modelling and mathematical analysis of certain problems arising from cell biology. We are particularly interested in cancer cells and their growth via several stages. To describe the event, multi-scale models are applied, involving continuously distributed environment variables and several components related to particles. Hybrid simulations are also carried out, using the discretization of environment variables and the Monte Carlo method for the principal particle variables. These modelling and simulation tools are put on rigorous mathematical foundations.

This monograph consists of four chapters. The first three chapters are concerned with modelling, while the last one is devoted to mathematical analysis. In Chap. 1, we examine molecular dynamics at the early stage of cancer invasion. We construct a pathway network model based on a biological scenario, and then determine its mathematical structures. In Chap. 2, we introduce the modelling over viewing several biological insights, using partial differential equations. Transport mechanics and movement via gradients are the main factors, and several models are introduced including the Keller–Segel systems. In Chap. 3, we turn to the method of averaging to model the movement of particles. This is based on mean field theories, using deterministic and stochastic approaches. Then, appropriate parameters for stochastic simulations are examined. The segment model is finally proposed as an application. In Chap. 4, we examine thermodynamical features of these models and how these structures are applied in mathematical analysis, that is negative chemotaxis, parabolic systems with non-local term accounting for chemical reactions, mass-conservative reaction-diffusion systems, and competitive systems of chemotaxis. We conclude this monograph with the method of the weak scaling limit applied to the Smoluchowski–Poisson equation.

Recent developments of cell biology using mathematical modelling occur at a very fast pace. Among them are bone metabolism, drug resistance, and signal transmission. We hope to extend and summarize these studies in future work.

We thank Prof. Angela Stevens for proposing to publish this text. Thanks are also due to Mrs. Keiko Itano for kind help in preparing these notes and providing several figures. This work is supported by JST-CREST project and JSPS Core-to-Core program A, Advanced Research Networks and JSPS Kakenhi Grant Number 16H06576.

Toyonaka, Japan   Takashi Suzuki
May 2017

Contents

1 Molecular Dynamics   1
   1.1 Pathway Modeling   1
   1.2 Pathway Analysis   8

2 Amounting the Balance   13
   2.1 Keller-Segel Model   13
   2.2 Invasion Model   19
   2.3 Smoluchowski-ODE Systems   27
   2.4 Smoluchowski-Poisson Systems   32

3 Averaging Particle Movements   39
   3.1 Deterministic Theory   39
   3.2 Random Theory   48
   3.3 Stochastic Simulations   52
   3.4 Segment Model   57

4 Mathematical Analysis   65
   4.1 Negative Chemotaxis   65
   4.2 Parabolic Systems with Non-local Term   78
   4.3 Reaction-Diffusion Systems   91
   4.4 Competitive System of Chemotaxis 106
   4.5 Method of the Weak Scaling Limit 114

Bibliography 135

Index 143

Chapter 1
Molecular Dynamics

Tumors are formed by abnormal growth of tissue, and cancer is the term used for malignant tumors. Cancer is a leading cause of death worldwide, accounting for 7.6 million deaths (around 13% of all deaths). Lung, stomach, liver, colon and breast cancer cause the most cancer deaths each year, and about 13.1 million deaths in 2030 are predicted by the International Agency for Research on Cancer (IARC). Here we describe a mathematical modeling which arose from a biological scenario for activation of basal membrane degradation at the early stage of cancer cell invasion.

1.1 Pathway Modeling

MT1-MMP (Membrane type-1 matrix metalloproteinase) is a protease working as an invasion apparatus of cancer cells. It is observed at the invasion front of cancer cells in the form of so-called invadopodia. MT1-MMP first activates MMP2, a secreted basal membrane enzyme which degrades collagen IV in the basal membrane. After the basal membrane is degraded, MT1-MMP degrades collagen I, II, II and laminin 1 and 5 of the ECM (Fig. 1.1).

The process of MMP2 activation is illustrated in [111] based on biological evidence. First, the extracellular pro-MMP2 is recruited. To attach MT1-MMP to the plasma membrane it binds another molecule called TIMP2. MT1-MMP itself forms dimers. Once the MT1-MMP, TIMP2, pro-MMP2, MT1 MMP is formed, the latter NT1-MMP cuts the connection pro-MMP2 and TIMP2. This process is called shedding activates MMP2, forming secretive type basement membrane degrading protease.

Henceforth we write a, b, and c for MMP2, TIMP2, and MT1-MMP, respectively. The above scenario indicates that the molecule $abcc$ is the origin to activate the secretive type basal membrane protease. Since this molecule is created only in the presence of b, it must be provided in sufficient amount at the begining. If the amount is too much, however, another molecule $abccba$ will be produced mainly. Hence

© Springer Nature Singapore Pte Ltd. 2017
T. Suzuki, *Mathematical Methods for Cancer Evolution*,
Lecture Notes on Mathematical Modelling in the Life Sciences,
DOI 10.1007/978-981-10-3671-2_1

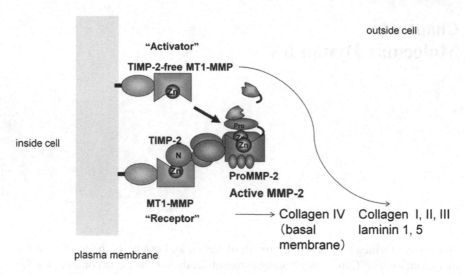

Fig. 1.1 ECM degradation

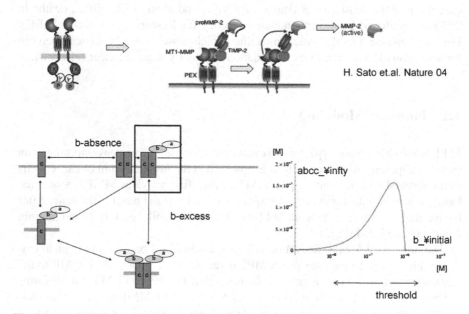

Fig. 1.2 a = MMP2, b = TIMP2, c = MT1-MMP

$b_0 - abcc_\infty$ curve will exhibit one peak, where b_0 and $abcc_\infty$ denote the initial and final concentrations of b and $abcc$, respectively (Fig. 1.2). Such a biological observation was used in [59] for the construction of a mathematical model, but below we shall adopt an axiomatic approach.

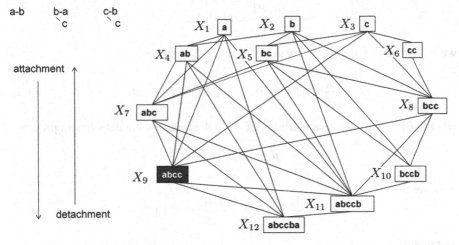

Fig. 1.3 Axiomatic model

The above biological model is concerned with the attachment and detachment of three kinds of molecule, a, b, and c. Let us explain rule of their polymerization. First, a has a hand which can attach b. Next, b has a hand that can attach a. This b has also a hand that can attach c. Finally, c has a hand that can attach b and also has another hand that can attach c. Excluding the other possibilities, we have 9 compounds made by a, b, and c (Fig. 1.3). We assume that these chemical reactions are subject to the law of *mass action*.

In this law, first, the *reaction rate* is defined as the production ratio due to chemical reaction in iso thermal liquid solution per unit time. By the definition, it is proportional to the frequency of molecular collisions. From experimental data, next, this frequency is proportional to the multiplication of the concentration of the reacting chemical substances. The law of mass action arises under this agreement, and hence the reaction rate is equal to a constant times the multiplication of concentrations of reacting chemical substances.

In the fundamental attachment process

$$A + B \rightarrow P \quad (k), \tag{1.1}$$

for example, it holds that

$$\frac{d[A]}{dt} = -k[A][B], \quad \frac{d[B]}{dt} = -k[A][B],$$

where $[A]$, $[B]$, and k denote the concentration of A, that of B, and the reaction rate, respectively. It holds also that

$$\frac{d[P]}{dt} = k[A][B],$$

which implies the mass conservations of A and B particles,

$$\frac{d}{dt}([A] + [P]) = \frac{d}{dt}([B] + [P]) = 0 \tag{1.2}$$

where $[P]$ denotes the concentration of P. In the fundamental detachment process

$$P \rightarrow A + B \quad (\ell) \tag{1.3}$$

we have

$$\frac{d[A]}{dt} = \ell[P], \quad \frac{d[B]}{dt} = \ell[P],$$

and hence

$$\frac{d[P]}{dt} = -\ell[P],$$

where ℓ denotes the reaction rate. Then (1.2) follows similarly. For the process combined with (1.1) and (1.3),

$$A + B \rightarrow AB \ (k), \quad AB \rightarrow A + B \ (\ell)$$

so we obtain

$$\frac{d[A]}{dt} = -k[A][B] + \ell[AB], \quad \frac{d[B]}{dt} = -k[A][B] + \ell[AB],$$
$$\frac{d[AB]}{dt} = k[A][B] - \ell[AB].$$

The fundamental processes (1.1) and (1.3) may be applied to modified molecules, for example,

$$AB + B \rightarrow BAB \quad (k), \quad BAB \rightarrow AB + B \quad (\ell). \tag{1.4}$$

Then we take

$$\frac{d[AB]}{dt} = -k[AB][B] + \ell[BAB], \quad \frac{d[B]}{dt} = -k[AB][B] + \ell[BAB],$$
$$\frac{d[BAB]}{dt} = k[AB][B] - \ell[BAB]$$

which guarantees the total mass conservations for A and B, that is,

$$\frac{d}{dt}([AB] + [BAB]) = 0, \quad \frac{d}{dt}([B] + [AB] + 2[BAB]) = 0. \tag{1.5}$$

Since the collision of particles induces chemical reactions with a definite rate, the mass action law is based on the assumption that the number of collisions of particles in a vessel of unit volume and per unit time is proportional to the product of the concentrations of these chemicals. Hence under (1.1) and (1.3) it is natural to assume that

$$A + BB \to ABB \ (2k), \quad ABB \to A + BB \ (\ell)$$

and

$$A + AB \to ABA \ (k), \quad ABA \to A + AB \ (2\ell).$$

This implies

$$\frac{d[A]}{dt} = -2k[A][BB] + \ell[ABB], \quad \frac{d[BB]}{dt} = -2k[A][BB] + \ell[ABB],$$

$$\frac{d[ABB]}{dt} = 2k[A][BB] - \ell[ABB]$$

and

$$\frac{d[A]}{dt} = -k[A][AB] + 2\ell[ABA], \quad \frac{d[AB]}{dt} = -k[A][AB] + 2\ell[ABA],$$

$$\frac{d[ABA]}{dt} = k[A][BB] - 2\ell[ABA],$$

respectively.

Considering

$$A + AB \to AAB \ (k), \quad AAB \to A + AB \ (\ell), \tag{1.6}$$

we have

$$\frac{d[A]}{dt} = -k[A][AB] + \ell[AAB], \quad \frac{d[AB]}{dt} = -k[A][AB] + \ell[AAB],$$

$$\frac{d[AAB]}{dt} = k[A][AB] - \ell[AAB],$$

which implies that

$$\frac{d}{dt}([A] + [AB] + 2[AAB]) = 0, \quad \frac{d}{dt}([AB] + [AAB]) = 0.$$

If N_A denotes the number of A molecules in a vessel of unit volume, the number of AA collision there per unit time is $N_A(N_A - 1)/2 \approx N_A^2/2$. Hence (1.6) implies

$$A + A \to AA \ (k/2), \quad AA \to A + A \ (\ell).$$

Then one obtains the system

$$\frac{d[A]}{dt} = 2\left(-\frac{k}{2}[A]^2 + \ell[AA]\right), \quad \frac{d[AA]}{dt} = \frac{k}{2}[A]^2 - \ell[AA]. \tag{1.7}$$

Here we took the sum of two equations to derive the first equation of (1.7). Then it the mass conservation of A particles holds:

$$\frac{d}{dt}([A] + 2[AA]) = 0.$$

Turning to the MT1-MMP model, we define the reaction rates k_1, k_2, k_3, ℓ_1, ℓ_2, and ℓ_3 by

$$
\begin{aligned}
a + b \rightarrow ab \ (k_1), \quad & ab \rightarrow a + b \ (\ell_1) \\
b + c \rightarrow bc \ (k_2), \quad & bc \rightarrow b + c \ (\ell_2) \\
c + c \rightarrow cc \ (k_3/2), \quad & cc \rightarrow c + c \ (\ell_3).
\end{aligned}
\tag{1.8}
$$

We have experimental data of these values, but we do not have any experimental data for the other reaction rates. Here we assume the rates in (1.8) for the reaction of modificated molecules, for example,

$$ab + c \rightarrow abc \ (k_2), \quad abc \rightarrow ab + c \ (\ell_2).$$

Hence we take

$$
\begin{aligned}
\frac{dX_3}{dt} &= -k_3 X_3 X_7 + \ell_3 X_9, \\
\frac{dX_7}{dt} &= -k_3 X_3 X_7 + \ell_3 X_9, \\
\frac{dX_9}{dt} &= k_3 X_3 X_7 - \ell_3 X_9
\end{aligned}
$$

for $abc + c \rightarrow abcc \ (k_3)$, $abcc \rightarrow abc + c \ (\ell_3)$, where $X_3 = [c]$, $X_7 = [abc]$, and $X_9 = [abcc]$. Similarly, we use

$$
\begin{aligned}
\frac{dX_3}{dt} &= -k_3 X_3 X_5 + \ell_3 X_8, \\
\frac{dX_5}{dt} &= -k_3 X_3 X_5 + \ell_3 X_8, \\
\frac{dX_8}{dt} &= k_3 X_3 X_5 - \ell_3 X_8
\end{aligned}
$$

for $bc + c \rightarrow bcc \ (k_3)$, $bcc \rightarrow bc + c \ (\ell_3)$, where $X_5 = [bc]$, $X_8 = [bcc]$, and also

$$\frac{dX_5}{dt} = -k_3 X_5 X_7 + \ell_3 X_{11},$$

$$\frac{dX_7}{dt} = -k_3 X_5 X_7 + \ell_3 X_{11},$$

$$\frac{dX_{11}}{dt} = k_3 X_5 X_7 - \ell_3 X_{11}$$

for $abc + bc \rightarrow abccb$ (k_3), $abccb \rightarrow abc + bc$ (ℓ_3), where $X_{11} = [abccb]$, and so forth.

Based on the diagram described in Fig. 1.3, we end up with

$$\frac{dX_1}{dt} = -k_1 X_1 X_2 - k_1 X_1 X_5 - k_1 X_1 X_8 - 2k_1 X_1 X_{10} - k_1 X_1 X_{11},$$

$$\frac{dX_2}{dt} = -k_1 X_1 X_2 - k_2 X_2 X_3 - 2k_2 X_2 X_6 - k_2 X_2 X_8 - k_2 X_2 X_9$$
$$+\ell_2 X_5 + \ell_2 X_8 + 2\ell_2 X_{10} + \ell_2 X_{11},$$

$$\frac{dX_3}{dt} = -k_2 X_2 X_3 - k_3 X_3 X_3 - k_2 X_3 X_4 \quad k_3 X_3 X_5 - k_3 X_3 X_7$$
$$+\ell_2 X_5 + 2\ell_3 X_6 + \ell_2 X_7 + \ell_3 X_8 + \ell_3 X_9,$$

$$\frac{dX_4}{dt} = k_1 X_1 X_2 - k_2 X_3 X_4 - 2k_2 X_4 X_6 - k_2 X_4 X_8 - k_2 X_4 X_9$$
$$+\ell_2 X_7 + \ell_2 X_9 + \ell_2 X_{11} + 2\ell_2 X_{12},$$

$$\frac{dX_5}{dt} = -k_1 X_1 X_5 + k_2 X_2 X_3 - k_3 X_3 X_5 - k_3 X_5 X_5 - k_3 X_5 X_7$$
$$-\ell_2 X_5 + \ell_3 X_8 + 2\ell_3 X_{10} + \ell_3 X_{11},$$

$$\frac{dX_6}{dt} = 2k_3 X_3 X_3 - 2k_2 X_2 X_6 - 2k_2 X_4 X_6 - \ell_3 X_6 + \ell_2 X_8 + \ell_2 X_9,$$

$$\frac{dX_7}{dt} = k_1 X_1 X_5 + k_2 X_3 X_4 - k_3 X_3 X_7 - k_3 X_5 X_7 \quad k_3 X_7 X_7$$
$$-\ell_2 X_7 + \ell_3 X_9 + \ell_3 X_{11} + 2\ell_3 X_{12},$$

$$\frac{dX_8}{dt} = -k_1 X_1 X_8 + k_2 X_2 X_6 - k_2 X_2 X_8 + k_3 X_3 X_5 - k_2 X_4 X_8$$
$$-\ell_2 X_8 + 2\ell_2 X_{10} + \ell_2 X_{11} - \ell_3 X_8,$$

$$\frac{dX_9}{dt} = k_1 X_1 X_8 - k_2 X_2 X_9 + k_3 X_3 X_7 + 2k_2 X_4 X_6 - k_2 X_4 X_9$$
$$-\ell_2 X_9 + \ell_2 X_{11} + 2\ell_2 X_{12} - \ell_3 X_9,$$

$$\frac{dX_{10}}{dt} = -2k_1 X_1 X_{10} + k_2 X_2 X_8 + 2k_3 X_5 X_5 - 2\ell_2 X_{10} - \ell_3 X_{10},$$

$$\frac{dX_{11}}{dt} = -k_1 X_1 X_{11} + 2k_1 X_1 X_{10} + k_2 X_2 X_9 + k_2 X_4 X_8 + k_3 X_5 X_7$$
$$-2\ell_2 X_{11} - \ell_3 X_{11},$$

$$\frac{dX_{12}}{dt} = k_1 X_1 X_{11} + k_2 X_4 X_9 + 2k_3 X_7 X_7 - 2\ell_2 X_{12} - \ell_3 X_{12}, \qquad (1.9)$$

where $X_1 = [a]$, $X_2 = [b]$, $X_3 = [c]$, $X_4 = [ab]$, $X_5 = [bc]$, $X_6 = [cc]$, $X_7 = [abc]$, $X_8 = [bcc]$, $X_9 = [abcc]$, $X_{10} = [bccb]$, $X_{11} = [abccb]$, and $X_{12} = [abccba]$. Finally, we use the experimental data [48, 99, 137] of reaction rates and initial concentrations of a, b, c for numerical simulations. This mathematical model is used in [48] for the study of cell biology.

1.2 Pathway Analysis

Mass action laws adapted above guarantee the total mass conservations of a, b, and c, so that the quantities

$$X_1 + X_4 + X_7 + X_9 + X_{11} + 2X_{12},$$
$$X_2 + X_4 + X_5 + X_7 + X_8 + X_9 + 2X_{10} + 2X_{11} + 2X_{12},$$
$$X_3 + X_5 + 2X_6 + X_7 + 2X_8 + 2X_9 + 2X_{10} + 2X_{11} + 2X_{12}$$

are invariant in time. This is actually confirmed by the model (1.9), since

$$\frac{d}{dt}(X_1 + X_4 + X_7 + X_9 + X_{11} + 2X_{12}) = 0,$$
$$\frac{d}{dt}(X_2 + X_4 + X_5 + X_7 + X_8 + X_9 + 2X_{10} + 2X_{11} + 2X_{12}) = 0,$$
$$\frac{d}{dt}(X_3 + X_5 + 2X_6 + X_7 + 2X_8 + 2X_9 + 2X_{10} + 2X_{11} + 2X_{12}) = 0.$$

The next observation is that the reaction rates (1.8) are used for all other processes. According to this assumption, all the paths are classified in three categories: the attachment and detachment of a-b, b-c, and c-c. Since $\ell_1 = 0$, the first case is reduced to

$$a + bB \rightarrow abB \quad (k_1),$$

where bB stands for all the compounds of $b's$. This process is listed as

$$X_1 + X_2 \rightarrow X_4, \quad X_1 + X_5 \rightarrow X_7, \quad X_1 + X_8 \rightarrow X_9,$$
$$X_1 + X_{10} \rightarrow X_{11}, \quad X_1 + X_{11} \rightarrow X_{12},$$

or overall,

$$X_1 + (X_2 + X_5 + X_8 + X_{10} + X_{11})$$
$$\rightarrow X_4 + X_7 + X_9 + X_{11} + X_{12} \quad (k_1). \tag{1.10}$$

It follows that

$$\frac{d}{dt}X_1 = -k_1 X_1(X_2 + X_5 + X_8 + X_{10} + X_{11}),$$

$$\frac{d}{dt}(X_2 + X_5 + X_8 + 2X_{10} + X_{11}) = -k_1 X_1(X_2 + X_5 + X_8$$
$$+ X_{10} + X_{11}). \tag{1.11}$$

The first and the second equations of (1.11) govern the conservations of mass of a and b, respectively, in this reaction. Since two b's of X_{10} are active in (1.10) we put X_{10} twice on the left-hand side of the second equation of (1.11). In fact the conservation of mass of b now reads

$$\frac{dX_1}{dt} = \frac{d}{dt}(X_2 + X_5 + X_8 + 2X_{10} + X_{11}).$$

The reactions

$$bB + cC \rightarrow BbcC \quad (k_2), \quad BbcC \rightarrow bB + cC \quad (\ell_2),$$

next, are listed as

$$X_2 + X_3 \leftrightarrow X_5, \quad X_2 + X_6 \leftrightarrow X_8,$$
$$X_2 + X_8 \leftrightarrow X_{10}, \quad X_2 + X_9 \leftrightarrow X_{11}, \tag{1.12}$$

and

$$X_4 + X_3 \leftrightarrow X_7, \quad X_4 + X_6 \leftrightarrow X_9,$$
$$X_4 + X_8 \leftrightarrow X_{11}, \quad X_4 + X_9 \leftrightarrow X_{12}, \tag{1.13}$$

or

$$(X_2 + X_4) + (X_3 + X_6 + X_8 + X_9)$$
$$\rightarrow X_5 + X_7 + X_8 + X_9 + X_{10} + X_{11} + X_{12} \quad (k_2),$$
$$X_5 + X_7 + X_8 + X_9 + X_{10} + X_{11} + X_{12}$$
$$\rightarrow (X_2 + X_4) + (X_3 + X_6 + X_8 + X_9) \quad (\ell_2). \tag{1.14}$$

Then we obtain

$$\frac{d}{dt}(X_2 + X_4) = -k_2(X_2 + X_4)(X_3 + X_6 + X_8 + X_9)$$
$$+ \ell_2(X_5 + X_7 + X_8 + X_9 + X_{10} + 2X_{11} + X_{12}),$$

$$\frac{d}{dt}(X_3 + 2X_6 + X_8 + X_9) = -k_2(X_2 + X_4)(X_3 + X_6 + X_8 + X_9)$$
$$+ \ell_2(X_5 + X_7 + X_8 + X_9 + X_{10} + 2X_{11} + X_{12}). \tag{1.15}$$

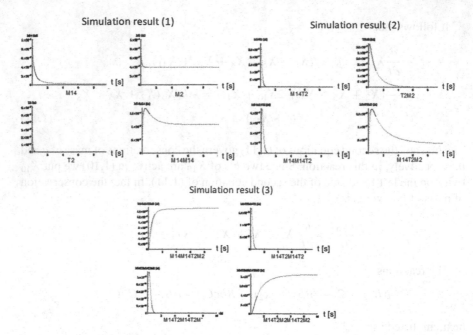

Fig. 1.4 Time course simulations

The first and the second equations of (1.15) govern the conservation of mass of b and c, respectively, in this reaction. Here the term $2X_6$ on the left-hand side of the second equation indicates that two c's of X_6 are active in the attachement of (1.14). The presence of the term $2X_{11}$ on the right-hand sides of both equations of (1.15) is due to the fact that X_{11} is involved by the reactions of both X_2 and X_4 in (1.12)–(1.13).

The reactions

$$cC_1 + cC_2 \rightarrow C_1ccC_2 \ (k_3), \quad C_1ccC_2 \rightarrow C_1c + C_2c \ (\ell_3),$$

finally, are listed as

$$X_3 + X_3 \leftrightarrow X_6, \quad X_3 + X_5 \leftrightarrow X_8, \quad X_3 + X_7 \leftrightarrow X_9, \tag{1.16}$$

$$X_5 + X_3 \leftrightarrow X_8, \quad X_5 + X_5 \leftrightarrow X_{10}, \quad X_5 + X_7 \leftrightarrow X_{11}, \tag{1.17}$$

and

$$X_7 + X_3 \leftrightarrow X_9, \quad X_7 + X_5 \leftrightarrow X_{11}, \quad X_7 + X_7 \leftrightarrow X_{12}. \tag{1.18}$$

A direct calculation actually implies

$$\frac{d}{dt}(X_3 + X_5 + X_7)$$
$$= -k_3\{(X_3 + X_5 + X_7)^2 + X_3^2 + X_5^2 + X_7^2\}$$
$$+2\ell_3(2X_6 + 2X_8 + 2X_9 + 2X_{10} + 2X_{11} + 2X_{12}). \qquad (1.19)$$

Using the above mentioned mass conservation and chemical reaction laws, we can decompose the system into several modules, in which the solution is represented explicitly [64] (Fig. 1.4).

Chapter 2
Amounting the Balance

Three factors: cell deformation, ECM degradation, and adhesion regulation arise at sub-cell level in the early stage of invasion of cancer cells. To understand the relations between them, the study of cell biology on proteins is integrated into mathematical models. Here we focus on *invadopodia*, formed on the surface of malignant tumor cells when they gain motility. Invadopodia are spiky, contain a lot of MT1-MMPs inside, and act as drills toward ECM. *Actin* inside a cell, on the other hand, takes two phases, F (solid) and G (liquid). The F-actin forms a network which casts a skeleton of the cell. Finally, a positive feedback loop is observed concerning up-regulation of MMPs. Our idea is to create in silico invadopodia using the above feedback loop and the switching fluctuations. Here we present the most fundamental tool of mathematical modeling, amounting the balance.

2.1 Keller-Segel Model

Mathematical formulas are used to describe several phenomena. For example, if $u = u(t)$ stands for the total amount of some species, then

$$u_t = \alpha$$

means that it is produced at the amount of α per unit time. The formula

$$u_t = ku,$$

is derived from $\alpha = ku$ which describes the case that this u produces itself with the rate k per unit time (Fig. 2.1). The negative sign arises similarly, that is,

© Springer Nature Singapore Pte Ltd. 2017
T. Suzuki, *Mathematical Methods for Cancer Evolution*,
Lecture Notes on Mathematical Modelling in the Life Sciences,
DOI 10.1007/978-981-10-3671-2_2

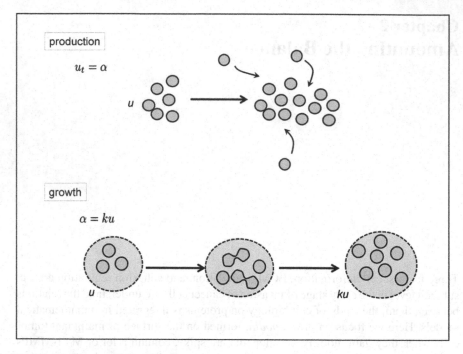

Fig. 2.1 Top down modeling (1)

$$v_t = -\beta$$

and

$$v_t = -\ell v.$$

If $u = u(x, t)$ denotes the density of some material at the position $x = (x_1, \ldots, x_n)$ and the time t, then

$$u_t = -\nabla \cdot j \tag{2.1}$$

formalizes the law of mass conservation, where j stands for the flux of u and

$$\nabla = \begin{pmatrix} \partial/\partial x_1 \\ \cdot \\ \cdot \\ \cdot \\ \partial/\partial x_n \end{pmatrix}$$

denotes the gradient operator (Fig. 2.2). Equation (2.1) arises together with the divergence formula,

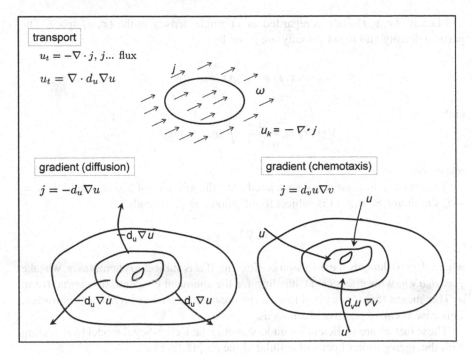

Fig. 2.2 Top down modeling (2)

$$\frac{d}{dt}\int_{\omega} u\, dx = -\int_{\partial\omega} \nu \cdot j\, dS, \qquad (2.2)$$

where ω is a domain with smooth boundary $\partial\omega$, and ν and dS denote the outer unit normal vector and surface element, respectively. The vector field $\nabla\varphi$ generated by the scalar field $\varphi = \varphi(x)$ points in direction where φ increases most and the length equal to its inclination. Equality (2.2) indicates the mass inside ω is lost with the amount of the outer normal component of j on $\partial\omega$.

From a microscopic point of view, Newton's equations of motion

$$\frac{dx}{dt} = v, \qquad m\frac{dv}{dt} = F \qquad (2.3)$$

represent the characteristic equations of the transport equation

$$\frac{\partial\rho}{\partial t} + \nabla_x\rho \cdot v + \frac{F}{m} \cdot \nabla_v\rho = 0. \qquad (2.4)$$

If $(x, v) = (x(t), v(t))$ and $\rho = \rho(x, v, t)$ are solutions to (2.3) and (2.4), respectively, then it holds that

$$\frac{d}{dt}\rho(x(t), v(t), t) = 0$$

and hence $\rho(x, v, t)dxdv$ is regarded as a particle density in the (x, v)-space. The particle density and mean velocity are given by

$$u(x, t) = \int \rho(x, v, t)dv$$

and

$$V(x, t) = \frac{1}{u(x, t)} \int \rho(x, v, t)vdv,$$

respectively.

The above j is sometimes associated with the gradient of a scalar field. If $j = -d_u \nabla u$, then $u = u(x, t)$ is subject to diffusion and (2.1) reads

$$u_t = \nabla \cdot d_u \nabla u,$$

where $d_u > 0$ denotes the diffusion coefficient. If u is subject to chemotaxis, we take $j = d_v u \nabla v$, with $v = v(x, t)$ standing for the chemical concentration attractive to u. This means that u is carried toward the area where v takes higher concentration. This case is called positive chemotaxis.

These factors are sufficient for understanding the Keller-Segel model [66] dealing with the aggregation of cells of cellular slime molds, that is,

$$\begin{aligned}
u_t &= \nabla \cdot (d_1(u, v)\nabla u) - \nabla \cdot (d_2(u, v)\nabla v), \\
v_t &= d_v \Delta v - k_1 vw + k_{-1} p + f(v)u, \\
w_t &= d_w \Delta w - k_1 vw + (k_{-1} + k_2)p + g(v, w)u, \\
p_t &= d_p \Delta p + k_1 vw - (k_{-1} + k_2)p \qquad \text{in } \Omega \times (0, T)
\end{aligned} \qquad (2.5)$$

with

$$d_1(u, v)\frac{\partial u}{\partial \nu} - d_2(u, v)\frac{\partial v}{\partial \nu} = 0,$$

$$\frac{\partial v}{\partial \nu} = \frac{\partial w}{\partial \nu} = \frac{\partial p}{\partial \nu} = 0 \qquad \text{on } \partial\Omega \times (0, T) \qquad (2.6)$$

where $\Omega \subset \mathbf{R}^N$ is a bounded domain with smooth boundary $\partial\Omega$. Here, $u = u(x, t)$, $v = v(x, t)$, $w = w(x, t)$, and $p = p(x, t)$ denote the density of the cellular slime mold, concentration of the chemical material, that of enzyme, and that of the complexes produced by v and w, respectively (Fig. 2.3). The nonlinearities $d_1(u, v)$, $d_2(u, v)$, $f(v)$, and $g(v, w)$ are introduced on biological grounds; a more quantitive based on biological experiments will be used if necessary.

Hence the first equation of (2.5) combined with the first boundary condition of (2.6) governs the transport of u with the null-flux boundary condition,

$$u_t = -\nabla \cdot j, \quad \nu \cdot j|_{\partial\Omega} = 0,$$

$u = u(x, t)$ cellular slime molds
$v = v(x, t)$ chemical
$w = w(x, t)$ enzyme
$p = p(x, t)$ complex

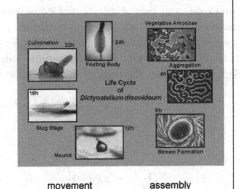

1. transport, gradient

(a) diffusion u, v, w, p

(b) chemotaxis $v \to u$

2. production $u \to (v, w)$

3. chemical reaction v, w, p

$$V + W \; \begin{matrix} (k_1) \to \\ \leftarrow (k_{-1}) \end{matrix} \; P \; \begin{matrix} (k_2) \\ \to \end{matrix} \; W + A$$

movement   assembly

Fig. 2.3 Dd - KS model http://www.zi.biologie.uni-muenchen.de/zoologie/diicty/dicty.html

where $j = -d_1(u, v)\nabla u + d_2(u, v)\nabla v$ stands for the flux of $u = u(x, t)$, composed of the diffusion and chemotaxis terms, $-d_1(u, v)\nabla u$ and $d_2(u, v)\nabla v$, respectively: (Fig. 2.2). The chemical materials v, w, and p are subject to the diffusion,

$$v_t = d_v \Delta v, \quad w_t = d_w \Delta w, \quad p_t = d_p \Delta p,$$

while the productions of v and w per unit time are propotional to u:

$$v_t = f(v)u, \quad w_t = g(v, w)u.$$

We are thus led to the system of ordinary differential equations

$$\begin{aligned} v_t &= -k_1 vw + k_{-1} p, \\ w_t &= -k_1 vw + (k_{-1} + k_2)p, \\ p_t &= k_1 vw - (k_{-1} + k_2)p, \end{aligned} \tag{2.7}$$

representing the mass action law for the chemical reaction

$$V + W \; \begin{matrix} (k_1) \to \\ \leftarrow (k_{-1}) \end{matrix} \; P \; \begin{matrix} (k_2) \\ \to \end{matrix} \; W + A. \tag{2.8}$$

We observe, however, several interactions through the organic hierarchy in the life, that is, individuals, organs, issues, cells, organelles, and molecular with different time scales. Othmer and Stevens [101] used functional relations, ordinary differential equations, and partial differential equations, to describe the signaling sensitivity, chemical reaction, and material transport, repectively. This model is applicable to several situations involving tissue, cell, sub-cell, and molecule levels, but, originally, describes the movement of one particle subject to a reinforced random walk (see Sect. 3.1). It takes the form of the Smoluchowski-ODE system

$$p_t = \nabla \cdot (D\nabla p - p\chi'(w)\nabla w),$$
$$w_t = g(p, w) \qquad\qquad \text{in } \Omega \times (0, T) \qquad\qquad (2.9)$$

with

$$D\frac{\partial p}{\partial \nu} - p\chi'(w)\frac{\partial w}{\partial \nu} = 0 \qquad\qquad \text{on } \partial\Omega \times (0, T) \qquad\qquad (2.10)$$

where p, w, D, χ', and g denote the existence probability, control species density, diffusion coefficient, chemotactic sensitivity, and growth factor, respectively (Fig. 2.4). The first equation of (2.9) is a generalization of the Smoluchowski equation and

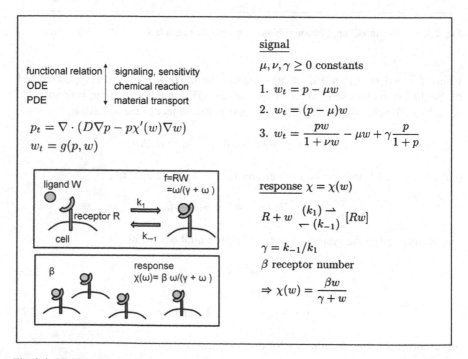

Fig. 2.4 Multi-scale model

(2.10) represents the null-flux condition. In the next chapter we shall see that it is based on modeling the movement of many particles.

The second equation of (2.9) is an ordinary differential equation. It governs the signaling process in accordance with the following functions:

1. (linear growth) $g(p, w) = p - \mu w$.
2. (exponential growth) $g(p, w) = (p - \mu)w$.
3. (saturated growth) $g(p, w) = \dfrac{pw}{1 + \nu w} - \mu w + \gamma \dfrac{p}{1 + p}$.

It is assumed that the control species w is subject to the process of ligand-receptor coupling

$$R + W \to F \ (k_1), \quad F \to R + W \ (k_{-1}).$$

It follows that

$$\frac{dr}{dt} = -k_1 r w + k_{-1} f,$$

$$\frac{dw}{dt} = -k_1 r w + k_{-1} f,$$

$$\frac{df}{dt} = k_1 r w - k_{-1} f,$$

where $r = [R]$, $w = [W]$, and $f = [F]$.

We have the total mass conservation $r + f = c$ and also $rw = \gamma f$ in the stationary state, where $\gamma = k_{-1}/k_1$ is the equilibrium constant. Putting $c = 1$, we obtain

$$f = \frac{w}{\gamma + w}.$$

Letting β be the receptor number, we now assume that

$$\chi(w) = \frac{\beta w}{\gamma + w},$$

that is, the sensitivity function $\chi = \chi(w)$ is proportional to the value of f.

2.2 Invasion Model

The Chaplain-Anderson model [18] concerning the invasion of cancer cells at the tissue level is formulated by

$$n_t = d_n \Delta n - \gamma \nabla \cdot (n \nabla c),$$
$$c_t = -\delta f c,$$
$$f_t = d_f \Delta f + \alpha n - \beta f \quad \text{in } \Omega \times (0, T), \tag{2.11}$$

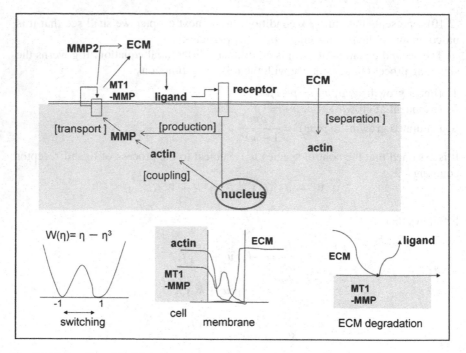

Fig. 2.5 Sub-cell model of invasion

with

$$d_n \frac{\partial n}{\partial \nu} - \gamma n \frac{\partial c}{\partial \nu} = \frac{\partial f}{\partial \nu} = 0 \quad \text{on } \partial\Omega \times (0, T). \tag{2.12}$$

Here, d_n, γ, δ, d_f, α, and β are positive constants, and n, c, and f denote the density of cells, that of ECM, and the concentration of MMP, respectively. In accordance with the Keller-Segel model (2.5), the chemical reaction is reduce to ODE. It is thus a multi-scale model and (n, c) is subject to the Smoluchowski ODE system (2.9) with (2.10). The unknown variable f stands for the concentration of enzyme, subject to diffusion, production by n, and self-decay.

More precisely, in the first equation of (2.11), the first term on the right-hand side describes the diffusive property of cancer cells at the tissue level. Then its second term reflects a feature called *haptotaxis*, which means that the cancer cells invade the ECM territory. Hence a linear haptotactic sensitivity is assumed. The second equation of (2.11) models the mass action associated with the ECM degradation by MMPs. The third equation of (2.11) describes the secretory and decay properties of MMPs produced by cancer cells.

A cell level model is used in [109]. First, the chemical reaction between f and c is considered. Using the density of ECM fragment, denoted by c_*, we take

$$c_t = -\kappa_c cf,$$

$$c_{*t} = d_{c_*}\Delta c_* + \kappa_c cf - \lambda_{c_*} c_* \quad \text{in } \Omega \times (0, T) \tag{2.13}$$

with

$$\frac{\partial c_*}{\partial \nu} = 0 \qquad \text{on } \partial\Omega \times (0, T), \tag{2.14}$$

because c_* is secretory. Here, κ_c, d_{c_*}, and λ_{c_*} are the reaction rate of $C + F \to C_*$, diffusion coefficient, and decay rate of c_*, respectively. Below we use similar notations. From the feedback loop, binding to the receptor of c_* provides the production of MMPs f and also the regulation of actin n inside the cell. These processes may be described by

$$n_t = d_n \Delta n - \gamma_n \nabla \cdot (n\nabla c_*),$$

$$f_t = d_f \Delta f + \kappa_f c_* - \lambda_f f \quad \text{in } \Omega \times (0, T), \tag{2.15}$$

with

$$d_n \frac{\partial n}{\partial \nu} - \gamma_n n \frac{\partial c_*}{\partial \nu} = \frac{\partial f}{\partial \nu} = 0 \quad \text{on } \partial\Omega \times (0, T). \tag{2.16}$$

There is, however, transport of f via n, which can be modeled by modifying the second equation of (2.15) and the corresponding boundary condition of (2.16) to

$$f_t = d_f \Delta f + \kappa_f c_* - \lambda_f f + \gamma_f \nabla \cdot (f\nabla n) \quad \text{in } \Omega \times (0, T)$$

and

$$d_f \frac{\partial f}{\partial \nu} + \gamma_f f \frac{\partial n}{\partial \nu} = 0 \qquad \text{on } \partial\Omega \times (0, T),$$

respectively. Thus n pushes f to the plasma membrane (see Fig. 2.5). Finally, since the events inside and outside the cell must be distinguished, we add the factor that ECM is repulsive to actin. This factor is treated by replacing the first equation of (2.15) by

$$n_t = d_n \Delta n + \nabla \cdot (n\nabla \chi(c)) - \gamma_n \nabla \cdot (n\nabla c_*).$$

Here, $\chi = \chi(c)$ is a suitable monotonically increasing function satisfying $\chi(c_k) = \infty$ with $0 < c_k - c_0 \ll 1$, where c_0 is a constant initially distributed for the value c outside the cell.

Thus we end up with the system

$$n_t = d_n \Delta n + \nabla \cdot (n\nabla \chi(c)) - \gamma_n \nabla \cdot (n\nabla c_*),$$

$$c_t = -\kappa_c cf,$$

$$c_{*t} = d_{c_*}\Delta c_* + \kappa_c cf - \lambda_{c_*} c_*,$$

$$f_t = d_f \Delta f + \kappa_f c_* - \beta f + \gamma_f \nabla \cdot (f\nabla n). \tag{2.17}$$

where d_n, d_{c_*}, d_f, γ_n, γ_f, κ_c, κ_f, and λ_{c_*} are constants (Fig. 2.5). The flux of n is now given by

$$j = -d_n \nabla n - n \nabla \chi(c) + \gamma_n n \nabla c_*,$$

involving diffusion, separation of n and c, and upregulate signaling from c_*. Then the boundary condition is given by

$$j \cdot \nu = \frac{\partial c_*}{\partial \nu} = \frac{\partial f}{\partial \nu} = 0.$$

The positive feedback loop model (2.17), however, does not cause the localization in space-time of the variables, n, f, and c. Since our motivation is to produce in silico invadopodia, here we assume that the fluctuation of the ECM degradation rate κ_c exhibits several peaks on the plasma membrane. In Fig. 2.5, the Allen-Cahn dynamics is used to describe such fluctuation, which, however, is not essential (see [109]).

In any case we obtain in silico invadopodia. Then the simulations of modified models, made by cutting some pathways, clarify the role of two feedback loops (Fig. 2.6), that is, ECM fragments to MMPs directly, and ECM fragments to MMP's indirectly via actin. Thus there are direct upregulation of f by c_* and also regulation of n by c_* inducing the transport of f. Numerical simulations then suggest that these feeback loops act as peaking and expansion of the plasma membrane, respectively.

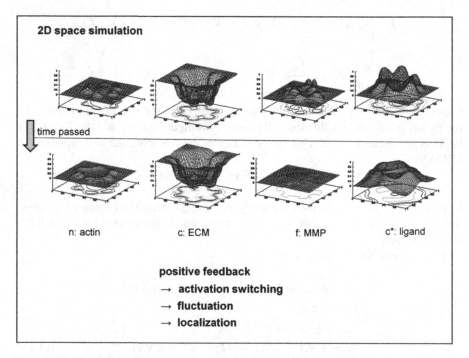

Fig. 2.6 In silico invadopodia

In spite of the presence of the phase separation term $n\nabla\chi(c)$ in the first equation of (2.17), in numerical simulations there arise the case that the region $n > 0$ becomes disconnected. Since the events taking place inside and outside the cell are different, treating the plasma membrane as a free boundary may be reasonable [37].

The plasma membrane can be represented by the level set like

$$\Gamma_t = \{x \in \Omega \mid \psi(x, t) = 0\}.$$

Here $\Omega \subset \mathbf{R}^N$ is the domain indicating the cancer cell and its enviroment, with smooth boundary $\partial\Omega$. The free boundary Γ_t is defined as the zero set of $\psi = \psi(x, t)$, assumed to be smooth. The interior and exterior the cell are denoted by ω_n^t and ω_c^t, respectively. Thus it follows that

$$\overline{\omega_n^t} \subset \Omega, \quad \partial\omega_n^t = \Gamma_t, \quad \omega_c^t = \Omega \setminus \overline{\omega_n^t}.$$

The velocity of the motion of Γ_t is the vector $v = v(x, t)$. This $v = v(x, t)$ is the gradient of the signal inside the cell denoted by σ. It is the driving force for the motion of the plasma membrane. Hence,

$$\psi_t + v \cdot \nabla\psi = 0, \quad v = \gamma_n \nabla\sigma \tag{2.18}$$

which implies

$$\frac{d}{dt}\psi(x(t), t) = \psi_t(x(t), t) + \frac{dx(t)}{dt} \cdot \nabla\psi(x(t), t) = 0$$

for $x = x(t)$ subject to

$$\frac{dx}{dt} = v(x, t).$$

The second idea is to restrict c_* and σ outside and inside the cell, respectively. Thus σ, n, and f are defined inside the cell, ω_n^t. There we assume that

$$
\begin{aligned}
n_t &= -\gamma_n \nabla \cdot (n\nabla\sigma), \\
f_t &= d_f \Delta f + \kappa_f \sigma + \gamma_f \nabla \cdot (f\nabla n) - \lambda_f f, \\
\sigma_t &= d_\sigma \Delta\sigma - \lambda_\sigma \sigma \qquad \qquad \text{in} \bigcup_{0<t<T} \omega_n^t \times \{t\},
\end{aligned}
\tag{2.19}
$$

with

$$d_f \frac{\partial f}{\partial\nu} + \gamma_f f \frac{\partial n}{\partial\nu} + fv \cdot \nu = 0,$$

$$\sigma = c_* \qquad \qquad \text{on} \bigcup_{0<t<T} \Gamma_t \times \{t\}, \tag{2.20}$$

recalling (2.17) (Fig. 2.7).

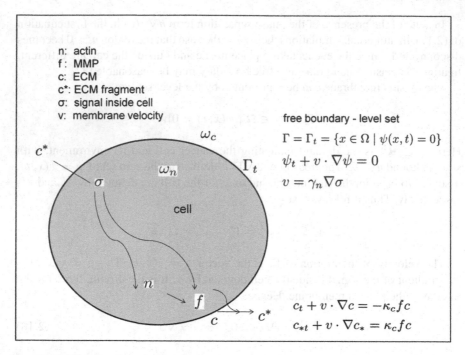

n: actin
f: MMP
c: ECM
c*: ECM fragment
σ: signal inside cell
v: membrane velocity

c^*

ω_c

Γ_t

ω_n

σ

cell

free boundary - level set

$$\Gamma = \Gamma_t = \{x \in \Omega \mid \psi(x, t) = 0\}$$

$$\psi_t + v \cdot \nabla \psi = 0$$

$$v = \gamma_n \nabla \sigma$$

n

f

c

c^*

$$c_t + v \cdot \nabla c = -\kappa_c f c$$

$$c_{*t} + v \cdot \nabla c_* = \kappa_c f c$$

Fig. 2.7 Individual cell model

Hence the diffusion term of the first equation of (2.19) is dropped, which implies that

$$n_t = -\nabla \cdot n v \qquad (2.21)$$

inside the cell. Then the total mass conservation of actin,

$$\frac{d}{dt} \int_{\omega_n^t} n \, dx = \int_{\omega_n^t} n_t \, dx + \int_{\partial \omega_n^t} n v \cdot \nu \, dS = 0$$

follows from the Liouville's formula for the first volume variation.

Given $v = v(x, t)$, the first equation of (2.18) is solved by the characteristic curve. Thus if the solution

$$\frac{dy}{dt} = v(y, t), \quad y|_{t=s} = x$$

is denoted by $y = U(t, s)x$, we have

$$\frac{d}{dt} \psi(U(t, 0)x, t) = 0.$$

Here we emphasize the group property of the propagator $\{U(t, s)\}$, described by

$$U(t, \tau) \circ U(\tau, s) = U(t, s), \quad U(t, t) = \text{Id}.$$

It follows that
$$\psi(x, t) = \psi_0(U(0, t)x),$$

where $\psi_0 = \psi(\cdot, 0)$. Equation (2.21) is solved similarly. In fact, it reads,

$$n_t + v \cdot \nabla n + n \nabla \cdot v = 0$$

and then it follows that

$$\frac{d}{dt} n(U(t, 0)x, t) = -n(U(t, 0)x, t)(\nabla \cdot v)(U(t, 0)x, t).$$

We thus obtain

$$n(U(t, 0)x, t) = n_0(x) \exp\left(-\int_0^t (\nabla \cdot v)(U(t', 0)x, t')dt'\right)$$

by the method of variation of constants, and therefore,

$$n(x, t) = n_0(U(0, t)x) \exp\left(-\int_0^t (\nabla \cdot v)(U(t', t)x, t')dt'\right), \qquad (2.22)$$

where $n_0 = n(\cdot, 0)$.

The second equation of (2.19) is involved by the conservation law,

$$\tilde{f}_t = -\nabla \cdot j, \quad j = -d_f \nabla f - \gamma_f f \nabla n.$$

Then the boundary condition for f of (2.20) guarantees the total mass conservation of $\tilde{f}$, in the sense that

$$\frac{d}{dt} \int_{\omega_n^i} \tilde{f}\, dx = \int_{\partial \omega_n^i} \left(-\nu \cdot j + f v \cdot \nu\right) dS = 0,$$

again by Liouville's formula for the first volume variation, where dS denotes the surface element.

The chemical reaction $c + f \to c_*$ is supposed to take place only on the boundary Γ_t. Using the material derivative

$$\frac{D}{Dt} = \frac{\partial}{\partial t} + v \cdot \nabla,$$

we obtain

$$c_t + v \cdot \nabla c = -\kappa_c f c,$$
$$c_{*t} + v \cdot \nabla c_* = \kappa_c f c \quad \text{on} \bigcup_{0<t<T} \Gamma_t \times \{t\}. \tag{2.23}$$

Then, similarly to (2.22), it follows that

$$c(x, t) = c_0(U(0, t)x) \exp\left(-\kappa_c \int_0^t f(U(t', t)x, t')dt'\right),$$

$$c_*(x, t) = c_{*0}(U(0, t)x) \exp\left(\kappa_c \int_0^t f(U(t', t)x, t')dt'\right) \quad \text{on} \bigcup_{0<t<T} \Gamma_t \times \{t\}$$

and total mass conservation holds in the sense that

$$\frac{D}{Dt}(c + c_*) = 0.$$

Outside the cell, we only use the diffusion and decay of c_*. Since its values on the free boundary are determined by the second equation in (2.23), we take

$$c_{*t} = d_{c_*} \Delta c_* - \lambda_{c_*} c_* \quad \text{in} \bigcup_{0<t<T} \omega_c^t \times \{t\}, \tag{2.24}$$

with

$$\frac{\partial c_*}{\partial \nu} = 0 \quad \text{on } \partial\Omega \times (0, T). \tag{2.25}$$

The model is now built up with (2.18), (2.19), (2.20), (2.23), (2.24), and (2.25). Then we obtain

$$\frac{d}{dt} \int_{\Gamma_t} c_* \, dS = \int_{\Gamma_*} (c_{*t} + (\nabla \cdot c_* \nu)v \cdot \nu) \, dS \tag{2.26}$$

$$= \int_{\Gamma_t} \left(c_{*t} + [(\nabla \cdot \nu)c_* + \frac{\partial c_*}{\partial \nu}]v \cdot \nu\right) dS$$

$$= \int_{\Gamma_t} \left(\kappa_c f c - v_\tau \frac{\partial c_*}{\partial \tau} + (\nabla \cdot \nu)c_* v \cdot \nu\right) dS, \tag{2.27}$$

by Liouville's formula for the first area variation. Here, the outer normal vector ν is taken, regarding $\omega_c = \omega_c^t$ as the interior. If $\omega_n = \omega_n^t$ has a peak denoted by Q, then $\nabla \cdot \nu < 0$ there. In the case that v is orthogonal to $\partial\omega_n$, the production of c_* is large at Q.

This model has the advantage of allowing simulations for many cells provided with the adhesion regulation (Fig. 2.8) see [37].

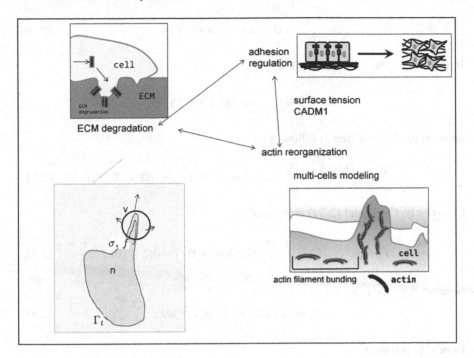

Fig. 2.8 Cell deformation

2.3 Smoluchowski-ODE Systems

Here we examine the mathematical validity of the multi-scale model described above, taking the simple form

$$q_t = \nabla \cdot (\nabla q - q \nabla \varphi(v)), \ v_t = q \quad \text{in } \Omega \times (0, T),$$
$$\frac{\partial q}{\partial \nu} - q \frac{\partial}{\partial \nu} \varphi(v) = 0 \quad \text{on } \partial\Omega \times (0, T),$$
$$q|_{t=0} = q_0(x) \geq 0, \ v|_{t=0} = v_0(x) \quad \text{in } \Omega, \tag{2.28}$$

where $\Omega \subset \mathbf{R}^N$ is a bounded domain with smooth boundary $\partial\Omega$, ν is the unit outer normal vector, $q_0 = q_0(x) > 0$ and $v_0 = v_0(x)$ are smooth functions of $x \in \overline{\Omega}$, and $\varphi : \mathbf{R} \to \mathbf{R}$ is a smooth function. Below we examine several multi-scale models of (2.9)–(2.10) via (2.28).

Imposing the condition

$$\frac{\partial v_0}{\partial \nu} = 0 \quad \text{on } \partial\Omega \tag{2.29}$$

to the initial value, we can replace the boundary condition by

$$\frac{\partial q}{\partial \nu} = 0 \quad \text{on } \partial\Omega \times (0, T).$$ (2.30)

In fact, (2.30) and (2.29) imply that

$$\frac{\partial v}{\partial \nu} = 0 \quad \text{on } \partial\Omega \times (0, T),$$

because $v_t = q$, and then it follows that

$$\frac{\partial q}{\partial \nu} - q \frac{\partial \varphi(v)}{\partial \nu} = \frac{\partial q}{\partial \nu} - q\varphi'(v) \frac{\partial v}{\partial \nu} = 0 \quad \text{on } \partial\Omega \times (0, T).$$ (2.31)

Conversely, (2.31) and (2.29) imply that

$$\frac{\partial q}{\partial \nu}(\cdot, t) - [q\varphi'(v)](\cdot, t) \int_0^t \frac{\partial q}{\partial \nu}(\cdot, t')dt' = 0 \quad \text{on } \partial\Omega \times (0, T),$$ (2.32)

because

$$v(x, t) = v_0(x) + \int_0^t q(x, t')dt'.$$ (2.33)

Hence, (2.30) holds.

System (2.28) may look restrictive, compared with a variety of (2.9)–(2.10). For example, problem (2.1) of [150] with $\chi(w) = \log w$ and $g(p, w) = \beta p - \mu w$ does not reduce to (2.28). Some cases, however, are still reducible to (2.28). First,

$$w_t = g(p, w)$$

is transformed to

$$v_t = q$$

for $g(p, w)$ given by the expressions below, where $\mu > 0$ is a constant:

$$g(p, w) = (p - \mu)w, \ w > 0 \quad \Rightarrow \quad v = \log w, \ q = p - \mu,$$
$$g(p, w) = p(\mu - w), \ w < \mu \quad \Rightarrow \quad v = -\log(\mu - w), \ q = p,$$
$$g(p, w) = -pw, \ w > 0 \quad \Rightarrow \quad v = -\log w, \ q = p.$$

The model (2.9)–(2.10), used in [72–75], is thus reduced to (2.28) in the following cases:

$$\chi'(w) = \frac{a(\beta - \alpha)}{(w + \alpha)(w + \beta)}, \quad g(p, w) = pw, \quad w > 0$$

$$\Rightarrow \quad \varphi(v) = a \log \frac{e^v + \alpha}{e^v + \beta},$$ (2.34)

$$\chi'(w) = \frac{a(\beta - \alpha)}{(w + \alpha)(w + \beta)}, \quad g(p, w) = -pw, \quad w > 0$$

$$\Rightarrow \quad \varphi(v) = a \log \frac{e^{-v} + \alpha}{e^{-v} + \beta}. \tag{2.35}$$

The case of negative chemotaxis,

$$\varphi \in C^3(\mathbf{R}), \quad \varphi' \leq 0 \leq \varphi''. \tag{2.36}$$

is obtained for $\alpha \geq \beta, v \geq \frac{1}{2} \log(\alpha\beta)$ and $\beta \leq \alpha, v \leq \frac{1}{2} \log(\alpha\beta)$ in (2.34) and (2.35), respectively. The other examples used by [101] are the following, where $a > 0$ is a constant:

$$\chi'(w) = a < 0, \quad g(p, w) = p(\mu - w), \quad w < \mu,$$
$$\chi'(w) = -a/w, \quad g(p, w) = p(\mu - w), \quad v \geq -\log \mu,$$
$$\chi'(w) = -a/(1 + w)^2, \quad g(p, w) = pw, \quad w > 0, \quad v \geq 0,$$
$$\chi'(w) = a, \quad g(p, w) = -pw, \quad w > 0,$$
$$\chi'(w) = a/w, \quad g(p, w) = -pw, \quad w > 0,$$
$$\chi'(w) = a/(1 + w)^2, \quad g(p, w) = -pw, \quad w > 0, \quad v \geq 0.$$

Furthermore, system (2.28) with (2.36) is equivalent to the one studied by [22], that is,

$$n_t = \nabla \cdot (\nabla n - n\chi'(c)\nabla c), \quad n > 0,$$
$$c_t = -cn, \quad c > 0 \qquad \qquad \text{in } \Omega \times (0, T),$$
$$\frac{\partial n}{\partial \nu} - \chi'(c)\frac{\partial c}{\partial \nu} = 0 \qquad \qquad \text{on } \partial\Omega \times (0, T), \tag{2.37}$$

where $\chi = \chi(c)$ is a C^2-function satisfying

$$\chi'(c) \geq 0, \quad c\chi''(c) + \chi'(c) \geq 0. \tag{2.38}$$

In fact, putting $v = -\log c$ and $q = n$, we obtain (2.28), for $\varphi = \varphi(v)$ where

$$\varphi(v) = \chi(c), \quad v = -\log c.$$

Then (2.38) coincides with (2.36) (see also [23]).

In the next section we show the global-in-time existence of the solution to (2.28) and its ergodic property in the case where the space dimension is one. The argument follows [106] for $\varphi(v) = -v$, using the continuous embedding of [81], namely,

$$L^2(0, T; H^1(\Omega)) \cap L^\infty(0, T; L^2(\Omega)) \hookrightarrow L^4(0, T; L^\infty(\Omega)).\qquad(2.39)$$

This result is a counterpart of the one obtained in [150] which says that if $\varphi(v) = v$, we have both global and blowup in finite time solutions, depending on their initial data. We note that $\varphi(v) = v$ does not satisfy $\varphi' \leq 0$.

The above property is valid also to the system

$$q_t = \nabla \cdot (\nabla q - q\nabla\varphi(v, w)),$$
$$v_t = q, \quad w_t = q \qquad\qquad\qquad\text{in } \Omega \times (0, T),$$
$$\frac{\partial q}{\partial \nu} = 0 \qquad\qquad\qquad\qquad\text{on } \partial\Omega \times (0, T),$$
$$q|_{t=0} = q_0, \quad v|_{t=0} = v_0, \quad w|_{t=0} = w_0 \quad\text{in } \Omega.\qquad(2.40)$$

Here we impose the compatibility condition

$$\frac{\partial v_0}{\partial \nu} = \frac{\partial w_0}{\partial \nu} = 0 \quad\text{on } \partial\Omega$$

which replaces the boundary condition by the zero-flux condition

$$\frac{\partial q}{\partial \nu} - q\frac{\partial\varphi(v, w)}{\partial \nu} = 0 \quad\text{on } \partial\Omega \times (0, T).$$

Then we obtain a similar result, assuming

$$\varphi = \varphi(v, w) \in C^3(\mathbf{R} \times \mathbf{R}),$$
$$\varphi_v, \varphi_w \leq 0, \varphi_{vv}, \varphi_{ww} \geq 0, \varphi_{vw} = 0.\qquad(2.41)$$

System (2.40) can describe several models.

The first example, found in [4], is modeling tumour induced angiogenesis. The variables in it are the endothelial cell density per unit area, denoted by n, the TAF (tumour angiogenesis factors) concentration f, and the matrix macromolecule fibronectin concentration c. Thus we have

$$n_t = D\Delta n - \nabla \cdot (\chi'(c)n\nabla c) - \rho_0\nabla \cdot (n\nabla f),$$
$$f_t = \beta n - \mu n f,$$
$$c_t = -\gamma n c,\qquad(2.42)$$

with

$$\chi'(c) = \frac{\chi_0}{1 + \alpha c}\qquad(2.43)$$

where $D, \rho_0, \beta, \mu, \gamma, \chi_0, \alpha > 0$ are constants. We can write (2.42) with initial and boundary conditions as

$$
\begin{aligned}
n_t &= \nabla \cdot (D\nabla n - n\nabla \log \Phi(c) - n\nabla \log \Psi(f)), \\
f_t &= \beta n - \mu n f, \quad c_t = -\gamma n c && \text{in } \Omega \times (0, T), \\
&(D\nabla n - n\nabla \log \Phi(c) - n\nabla \log \Psi(f)) \cdot \nu = 0 && \text{on } \partial\Omega \times (0, T), \\
n|_{t=0} &= n_0 > 0, \quad f|_{t=0} = f_0 > 0, \quad c|_{t=0} = c_0 > 0 && \text{in } \Omega.
\end{aligned}
\tag{2.44}
$$

Here, $\Phi, \Psi : \mathbf{R} \to \mathbf{R}$ are smooth positive functions satisfying

$$
\chi(c) = \log \Phi(c), \quad \rho_0 f = \log \Psi(f).
$$

Assuming

$$
f_0 > \frac{\beta}{\mu} \qquad \text{in } \Omega,
$$

$$
\frac{\partial n_0}{\partial \nu} = \frac{\partial c_0}{\partial \nu} = \frac{\partial f_0}{\partial \nu} = 0 \qquad \text{on } \partial\Omega,
\tag{2.45}
$$

we put $\tau = Dt$, $q = n$, $v = -\frac{D}{\gamma} \log c$, $w = -\frac{D}{\mu} \log(\mu f - \beta)$, and

$$
\begin{aligned}
\varphi(v, w) &= \log \tilde{\Phi}(v) + \log \tilde{\Psi}(w), \\
\tilde{\Phi}(v) &= \Phi(e^{-\gamma v/D})^{1/D}, \\
\tilde{\Psi}(w) &= \Psi(\mu^{-1}(\beta + e^{-\mu w/D}))^{1/D}.
\end{aligned}
$$

Then we obtain (2.40) from (2.44), writing t for τ, and are able to verify all the assumptions required for $\varphi = \varphi(v, w)$.

The same treatment is possible for the other model of angiogenesis used in [5], that is,

$$
\Phi(c) = e^{\varphi_0 c}, \qquad \Psi(f) = e^{\rho_0 f}.
$$

These models of angiogenesis are derived from several formulas in Sect. 2.1, accounting for the effects of angiogenesis, chemotaxis, and haptotaxis to fit the experimental data.

The system (2.28) is equivalent to the evolution equation with strong dissipation

$$
\begin{aligned}
v_{tt} &= \Delta v_t - \nabla \cdot (v_t \nabla \varphi(v)) && \text{in } \Omega \times (0, T), \\
\frac{\partial v}{\partial \nu} &= 0 && \text{on } \partial\Omega \times (0, T).
\end{aligned}
$$

This formulation is used to study the blowup and global-in-time solutions in [74, 80, 150]. For example, if the nonlinearity $\varphi = \varphi(v)$ is assumed to be bounded together with higher-order derivatives and to satisfy

$$
\lim_{v \uparrow +\infty} \varphi'(v) = 0,
$$

then a global-in-time solution exists, provided that

$$q_0(x) = \gamma + q_1(x), \quad \gamma \gg 1, \quad \int_\Omega q_1(x)\, dx = 0.$$

A similar result holds also for (2.42)–(2.43) (see [72, 73, 75]). Some models of [101] are reduced to taking $\varphi = \varphi(v, w)$ as in (2.34)–(2.35); these cases are studied by [74].

In the counterpart of (2.45),

$$0 < f_0 < \frac{\beta}{\mu},$$

there are a priori bounds of the solution to problem (2.42)–(2.43) under the assumption of

$$(\beta - \mu f_0)^{\gamma/\beta} \ll c_0.$$

Then a global-in-time solution exists and it converges to the stationary solution, by the comparison principle [35].

2.4 Smoluchowski-Poisson Systems

The chemical process (2.8) is sometimes replaced by the Michaelis-Menten process. It arises when one assumes that (w, p) is in the quasi-stationary state. Using the conservation of total mass $(w + p)_t = 0$, it is formulated as

$$k_1 v w - (k_{-1} + k_2) p = 0, \quad w + p = c, \tag{2.46}$$

where c is a constant. Then it holds that

$$w = \frac{(k_{-1} + k_2)c}{k_1 v + k_{-1} + k_2}, \quad p = \frac{ck_1 v}{k_1 v + k_{-1} + k_2},$$

and hence

$$-k_1 v w + k_{-1} p = -\frac{ck_1 k_2}{k_1 v + k_{-1} + k_2} v.$$

Equation (2.7) thus reduces to

$$v_t = -k(v)v, \quad k(v) = \frac{ck_1 k_2}{k_{-1} + k_2 + k_1 v},$$

and then the Keller-Segel system (2.5)–(2.6) implies

$$u_t = \nabla \cdot (d_1(u, v)\nabla u) - \nabla \cdot (d_2(u, v)\nabla v),$$
$$v_t = d_v \Delta v - k(v)v + f(v)u \qquad \text{in } \Omega \times (0, T), \qquad (2.47)$$

with

$$\frac{\partial u}{\partial \nu} = \frac{\partial v}{\partial \nu} = 0 \qquad\qquad \text{on } \partial\Omega \times (0, T). \qquad (2.48)$$

Nanjundiah [95] considered the case where $d_1(u, v)$, $k(v)$, and $f(v)$ are constants, and $d_2(u, v) = u\chi'(v)$ in (2.47)–(2.48) which leads to the problem

$$u_t = d_u \Delta u - \nabla \cdot (u\nabla\chi(v)),$$
$$v_t = d_v \Delta v - b_1 v + b_2 u \qquad \text{in } \Omega \times (0, T),$$
$$\frac{\partial u}{\partial \nu} = \frac{\partial v}{\partial \nu} = 0 \qquad\qquad \text{on } \partial\Omega \times (0, T). \qquad (2.49)$$

Here $\chi'(v)$ stands for the chemotactic sensitivity and henceforth $\chi = \chi(v)$ is called the sensitivity function.

Reducing the second equation to an ODE, we get the Smoluchowski-ODE system studied in (2.3),

$$q_t = \nabla \cdot (\nabla q - q\nabla\varphi(v))$$
$$v_t = q \qquad\qquad \text{in } \Omega \times (0, T)$$
$$\frac{\partial q}{\partial \nu} - q\frac{\partial\varphi(v)}{\partial \nu} = 0 \qquad \text{on } \partial\Omega \times (0, T). \qquad (2.50)$$

An asymptotic analysis, on the other hand, produces the simplified system for the linear sensitivity function $\chi(v) = \chi v$, the Smoluchowski-Poisson equation [20, 56]:

$$u_t = \nabla \cdot (\nabla u - u\nabla v),$$
$$-\Delta v = u - \frac{1}{|\Omega|}\int_\Omega u \, dx \qquad \text{in } \Omega \times (0, T), \qquad \int_\Omega v = 0,$$
$$\frac{\partial u}{\partial \nu} - u\frac{\partial v}{\partial \nu} = \frac{\partial v}{\partial \nu} = 0 \qquad \text{on } \partial\Omega \times (0, T). \qquad (2.51)$$

Still we can observe several mesoscopic key factors in (2.51) used for the original top-down modeling.

In fact, the first equation represents mass conservation law (2.15),

$$u_t = -\nabla \cdot j \qquad\qquad\qquad\qquad (2.52)$$

with the flux of u defined by

$$j = -\nabla u + u\nabla v. \qquad\qquad\qquad\qquad (2.53)$$

The null-flux boundary condition is imposed in (2.51), which guarantees the conservation of the total mass:

$$\frac{d}{dt} \int_\Omega u \, dx = 0. \tag{2.54}$$

In (2.53) the chemical substance v stands for the carrier of the cells u. The diffusion $-\nabla u$ is thus competing the chemotaxis $u\nabla v$ according to the phenomenological relation. The second equation of (2.51), on the other hand, describes a coarsed process of generation of the chemical potential ∇v from the particle density u using the Poisson equation. These features of (2.51) reflect several principles used in the bottom-up modeling. System (2.51) is actually obtained also by bottom-up modeling based on the movements of particles described in Sects. 3.1 and 3.2. The conservation law (2.52) thus reads

$$\frac{\partial u}{\partial t} + \nabla \cdot (uV) = 0,$$

with the transport velocity

$$V = u^{-1}j = -\nabla \log u + \nabla v.$$

The Poisson part

$$-\Delta v = u - \frac{1}{|\Omega|} \int_\Omega u \, dx \text{ in } \Omega, \quad \frac{\partial v}{\partial \nu} = 0 \text{ on } \partial\Omega, \quad \int_\Omega v \, dx = 0, \tag{2.55}$$

is uniquely solvable, which may be written as $v = (-\Delta)^{-1}u$ for simplicity. The strong maximum principle, on the other hand, guarantees that $u > 0$ in $\overline{\Omega} \times (0, T)$, provided that $u_0 \geq 0$ is not identically 0. Writing the Smoluchowski part as

$$u_t = \nabla \cdot u\nabla(\log u - v), \quad \left. \left(\frac{\partial u}{\partial \nu} - u\frac{\partial v}{\partial \nu} \right) \right|_{\partial\Omega} = 0, \tag{2.56}$$

we obtain

$$\int_\Omega u_t(\log u - v) \, dx = -\int_\Omega u|\nabla(\log u - v)|^2 \, dx, \tag{2.57}$$

with the left-hand side equal to

$$\frac{d}{dt} \left\{ \int_\Omega u(\log u - 1) \, dx - \frac{1}{2}\langle (-\Delta)^{-1}u, u \rangle \right\}.$$

This variational structure is physically justified by the fact that

$$\mathcal{F}(u) = \int_\Omega u(\log u - 1) \, dx - \frac{1}{2}\langle (-\Delta)^{-1}u, u \rangle$$

represents the Helmholtz free energy.

Using this model, it is rigorously shown in [56] that chemotaxis can serve as the main mechanism for the onset of self-organization of the *Dictyostelium discoideum* (Dd amoebae). Later studies are concerned with the blowup threshold [10, 36, 89, 90, 92, 118, 125], formation of collapse singularity [43, 117], mass quantization [43, 122, 128], sub-collapses which results in type II blowup rates [43, 94, 115], and the free energy transmission of the blowup solution [122, 123], while the blowup threshold does not arise for the full parabolic-parabolic system [11] (see [61, 145] for a related model).

Turning to the mesoscopic modeling, the phase field model arises with the chemical potential

$$\mu = \delta \mathcal{F}(\varphi),$$

where $\mathcal{F} = \mathcal{F}(\varphi)$ stands for the free energy $\mathcal{F} = \mathcal{F}(\varphi)$ defined by the order parameter $\varphi = \varphi(x)$. These equations are classified into models (A), (B), and (C) [40, 45]. The order parameter, $\varphi = \varphi(x, t)$, $(x, t) \in \Omega \times (0, T)$, represents the state of the material, while $\mathcal{F} = \mathcal{F}(\varphi)$ is a quantity determined by φ. Hence the system moves toward equilibrium: $\mathcal{F}(\varphi)$ decreases and attains a local minimum at the equilibrium. The chemical potential $\delta \mathcal{F}(\varphi)$ is defined by

$$\langle \psi, \delta \mathcal{F}(\varphi) \rangle = \frac{d}{ds} \mathcal{F}(\varphi + s\psi) \Big|_{s=0}, \tag{2.58}$$

where the pairing $\langle \ , \ \rangle$ is usually identified with the L^2 inner product.

The model (A) equation is formulated by

$$\varphi_t = -K \delta \mathcal{F}(\varphi) \quad \text{in } \Omega \times (0, T),$$

where K is a positive quantity, possibly associated with φ. Then it holds that

$$\frac{d}{dt} \mathcal{F}(\varphi) = -\int_\Omega K \delta \mathcal{F}(\varphi)^2 \leq 0.$$

The model (B) equation takes the form

$$\varphi_t = \nabla \cdot (K \nabla \delta \mathcal{F}(\varphi)) \quad \text{in } \Omega \times (0, T),$$

$$K \frac{\partial}{\partial \nu} \delta \mathcal{F}(\varphi) = 0 \quad \text{on } \partial\Omega \times (0, T).$$

In this case we obtain

$$\frac{d}{dt} \int_\Omega \varphi \, dx = \int_{\partial\Omega} K \frac{\partial}{\partial \nu} \delta \mathcal{F}(\varphi) \, dS = 0,$$

$$\frac{d}{dt} \mathcal{F}(\varphi) = -\int_\Omega K \, |\nabla \delta \mathcal{F}(\varphi)|^2 \, dx \leq 0.$$

The equations of models (A) and (B) describe thermodynamically closed and thermodynamically-materially closed systems, respectively. Neither of them admits non-trivial periodic-in-time solutions.

The stationary state is actually defined by the vanishing of the free-energy consumption, that is,

$$\frac{d}{ds}\mathcal{F}(\varphi + s\psi)\bigg|_{s=0} = 0, \quad \forall\psi, \tag{2.59}$$

and

$$\frac{d}{ds}\mathcal{F}(\varphi + s\psi)\bigg|_{s=0} = 0, \quad \forall\psi, \quad \int_\Omega \psi\, dx = 0, \quad \int_\Omega \varphi\, dx = \lambda, \tag{2.60}$$

respectively, in model (A) and (B) equations, where λ is a constant prescribed by the initial value. The linearized stability of the stationary state φ means, similarly, that

$$Q(\psi, \psi) \equiv \frac{1}{2}\frac{d^2}{ds^2}\mathcal{F}(\varphi + s\psi)\bigg|_{s=0} > 0, \quad \forall\psi \neq 0$$

and

$$Q(\psi, \psi) > 0 \quad \forall\psi \neq 0, \quad \int_\Omega \psi\, dx = 0$$

for the model (A) and (B) equations, respectively.

The Smoluchowski-Poisson equation (2.51) is the model (B) equation associated with the Helmholtz free energy. First, the Smoluchowski part describes mass conservation (2.52) with the flux given by (2.53). The chemotactic sensitivity v, on the other hand, is the solution to the Poisson part (2.55). Since $u = u(x, t)$ represents particle distribution function, this $v = v(x, t)$ acts as a potential of attractive chemotaxis. Here, (2.55) is equivalent to

$$v = \int_\Omega G(\cdot, x')u(x')\, dx' \tag{2.61}$$

with $G = G(x, x')$ denoting the Green function. Since

$$G(x, x') \approx \Gamma(x - x')$$

with

$$\Gamma(x) = \begin{cases} \dfrac{1}{4\pi} \cdot \dfrac{1}{|x|}, & N = 3 \\ \dfrac{1}{2\pi}\log\dfrac{1}{|x|}, & N = 2, \end{cases}$$

Eq. (2.55) may be regarded as the formation of gravitational field created by the particle density. The inner (potential) energy of this system, therefore, is defined by

$$E = -\frac{1}{2} \iint_{\Omega \times \Omega} G(x, x') u(x) u(x') \, dx dx'.$$

Here we take the minus sign because of the self-attractivity of gravitation. The factor $\frac{1}{2}$ appears as usual because Newton's third law of action-reaction induces the symmetry of the kernel in (2.61),

$$G(x', x) = G(x, x'), \tag{2.62}$$

Since u is the particle density, on the other hand, we introduce the entropy by

$$S = -\int_{\Omega} u(\log u - 1) \, dx.$$

Under the normalization $T = 1$, the Helmholtz free energy $F = E - TS$ thus takes the form

$$\mathcal{F}(u) = \int_{\Omega} u(\log u - 1) - \frac{1}{2} \iint_{\Omega \times \Omega} G(x, x') u(x) u(x') \, dx dx'. \tag{2.63}$$

The Smoluchowski-Poisson equation is an adiabatic limit of a kinetic equation with fluctuation and friction (see [143, 144] for more details). Equation (2.51) actually describes the motion of the mean field of many self-attracting particles subject to the second law of thermodynamics. Hence one has the decrease of free energy besides the total mass conservation (2.54). The phenomenological formulation of model (B) equation is consistent with the mesoscopic approach using kinetic equation in this case. In fact, the first variation $\delta \mathcal{F}(u)$ of $\mathcal{F}(u)$ is defined by

$$\langle w, \delta \mathcal{F}(u) \rangle = \frac{d}{ds} \mathcal{F}(u + sw) \Big|_{s=0}.$$

Identifying again this pairing $\langle \, , \, \rangle$ with the L^2 inner product, we obtain

$$\delta \mathcal{F}(u) = \log u - v, \quad v = \int_{\Omega} G(\cdot, x') u(x') \, dx'.$$

Thus (2.56) reads

$$u_t = \nabla \cdot (u \nabla \delta \mathcal{F}(u)) \qquad \text{in } \Omega \times (0, T),$$

$$u \frac{\partial}{\partial \nu} \delta \mathcal{F}(u) = 0 \qquad \text{on } \partial\Omega \times (0, T). \tag{2.64}$$

This is a form of the model (B) equation derived from the free energy $\mathcal{F} = \mathcal{F}(u)$. Consequently,

$$\frac{d}{dt} \int_\Omega u \, dx = - \int_{\partial\Omega} u \frac{\partial}{\partial\nu} \delta\mathcal{F}(u) \, dS = 0,$$

$$\frac{d}{dt} \mathcal{F}(u) = - \int_\Omega u \, |\nabla\delta\mathcal{F}(u)|^2 \, dx \le 0. \tag{2.65}$$

The second inequality in (2.65) means the decrease of free energy, while the first equality, combined with $u = u(x, t) \ge 0$, assures the total mass conservation

$$\lambda \equiv \|u_0\|_1 = \|u(\cdot, t)\|_1 , \qquad t \in [0, T_{\max}). \tag{2.66}$$

Relation (2.66) leads to the selection of the space dimension $N = 2$ for the formation of collapses, using dimensional analysis [20].

The stationary problem (2.60), on the other hand, is given by

$$\log u - v = \text{constant}, \quad v = (-\Delta)^{-1} u, \quad \int_\Omega u \, dx = \lambda, \tag{2.67}$$

where $v = (-\Delta)^{-1} u$ indicates (2.61). Then it holds that

$$-\Delta v = \lambda \left(\frac{e^v}{\int_\Omega e^v} - \frac{1}{|\Omega|} \right) \text{ in } \Omega, \quad \frac{\partial v}{\partial\nu} = 0 \text{ on } \partial\Omega, \quad \int_\Omega v \, dx = 0, \tag{2.68}$$

since

$$u = \frac{\lambda e^v}{\int_\Omega e^v dx}, \quad v = (-\Delta)^{-1} u.$$

The profile of the solution in (2.68) by [116] suggests the quantized blowup mechanism for (2.51). Later, this property was studied by many authors (see [122, 125, 128] and the references therein). See also Sect. 4.5.

Chapter 3
Averaging Particle Movements

Several formulas in Sect. 2.1 have microscopic backgrounds. For example, the three diffusion equations $u_t = \nabla \cdot d_u \nabla u$, $u_t = d_u \Delta u$, and $u_t = \Delta(d_u u)$ are not equivalent unless the diffusion coefficient d_u is spatially homogeneous. Actually they have different underlying microscopic structures (see [98]), and distinguishing them is required to improve the model. In the mean field theory, these formulas are derived by taking the limit of the distribution function of many particles. Several methods are proposed, based on deterministic or random points of view. They are useful even in practice, for example, in adjusting parameters of stochastic simulation. For instance, the fact that the average rate of a stochastic reaction should be equal to that of deterministic simulation tells us how the parameters should be selected and to what extent the simulation has picked up the fluctuations of molecules.

3.1 Deterministic Theory

The diffusion equation and its relatives such as the Smoluchowski equation in (2.9) arise as the mean-field limit of microscopic states representing the motion of particles. Transport theory uses the master equation representing mass conservation in the presence of jump processes. Two cases can be distinguished, the velocity jump and the space jump (see [100]). The work [101] is concerned with the latter case. It assumes that a particle walks on the lattice

$$\mathcal{Z} = \{\ldots, -n-1, -n, -n+1, \ldots, -1, 0, +1, \ldots, n-1, n, n+1, \ldots\}$$

with the probabilities $T_n^{\pm}$ for the transitions $n \mapsto n \pm 1$. If $p_n(t)$ denotes the particle density at the site n and the time t, then the corresponding master equation takes the form

© Springer Nature Singapore Pte Ltd. 2017
T. Suzuki, *Mathematical Methods for Cancer Evolution*,
Lecture Notes on Mathematical Modelling in the Life Sciences,
DOI 10.1007/978-981-10-3671-2_3

$$\frac{dp_n}{dt} = T_{n-1}^+ p_{n-1} + T_{n+1}^- p_{n+1} - (T_n^+ + T_n^-)p_n. \tag{3.1}$$

In the context of biology, it may be reasonable to assume that some other species are controling $T_n^\pm$. Several cases are examined according to the strategies of these control species [88, 98].

In the simplest case of local information, $T_n^\pm$ is a function of (x, t), $x = n\Delta x$, where Δx denotes the mesh size of the lattice. In the formal argument we assume the mean-field limit $T_n^\pm = T(n\Delta x, t)$ with the expected smooth function $T = T(x, t)$. Then it holds that

$$\frac{dp_n}{dt} = T_{n-1} p_{n-1} + T_{n+1} p_{n+1} - 2T_n p_n, \quad T_n = T(n\Delta x, t).$$

Using the three-point formula

$$f(x+h) + f(x-h) - 2f(x) = h^2 f''(x) + o(h^2), \quad h \to 0$$

and the renormalization $t' = t\Delta t$, we get the limit equation

$$\frac{\partial p}{\partial t'} = D\frac{\partial^2}{\partial x^2}(Tp),$$

under the assumption that

$$D = \frac{(\Delta x)^2}{\Delta t} \tag{3.2}$$

is a constant. The higher-dimensional case is similar and the limit equation

$$\frac{\partial p}{\partial t} = D\Delta(Tp) \tag{3.3}$$

arises, with t replacing t'.

In the other case, when

$$T_n^\pm = T_{n\pm1} = T((n \pm 1)\Delta x, t),$$

relation (3.1) reads

$$\begin{aligned}
\frac{dp_n}{dt} &= T_n p_{n-1} + T_n p_{n+1} - (T_{n+1} + T_{n-1})p_n \\
&= T_n(p_{n-1} + p_{n+1} - 2p_n) - (T_{n+1} + T_{n-1} - 2T_n)p_n,
\end{aligned}$$

which in the limit equation yields the cross diffusion

$$\frac{\partial p}{\partial t} = D(T\Delta p - (\Delta T)p). \tag{3.4}$$

In the barrier model, on the other hand, we assume

$$T_n^\pm = T_{n\pm1/2} = T((n\pm1/2)\Delta x, t).$$

Then it follows that

$$\begin{aligned}
\frac{dp_n}{dt} &= T_{n-1/2}p_{n-1} + T_{n+1/2}p_{n+1} - (T_{n+1/2} + T_{n-1/2})p_n \\
&= T_{n+1/2}(p_{n+1} - p_n) - T_{n-1/2}(p_n - p_{n-1}).
\end{aligned}$$

Therefore, the limit equation now takes the form

$$\frac{\partial p}{\partial t} = D\frac{\partial}{\partial x}\left(T\frac{\partial p}{\partial x}\right) \tag{3.5}$$

(see [101]).

Here we examine two aspects of this model. First, the coefficient D in (3.2) is to be determined by physical constants. In fact, it is the diffusion coefficient associated with Einstein's formula

$$\tau = \frac{(\Delta x)^2}{2ND}, \tag{3.6}$$

where τ, Δx, N, and D are the mean waiting time, mean jump length, space dimension, and diffusion coefficient, respectively. In [54] the constant D in (3.2) is taken to be consistent with (3.6), using a particle distribution function continuously defined, subject to a master equation provided with discrete particle jump processes in space and time. The next aspects concerns the formulation of collisions of particles arising in chemical reactions. Such a process is actually used in stochastic simulation of molecular dynamics, while two particles walking on lattices of their own do not meet each other in a lattice model. Hence we have to assume the same lattice for these molecules, which results in the same jump length. Particle densities defined continuously in space and time thus have a second advantage.

Let $q(x, t)$ be the particle density, defined for all $x \in \mathbf{R}^N$ and $t > 0$. We assume the simplest case, namely that the particle takes the constant jump length Δx. Thus let $T(x, t; \omega)$ and Δt be the transition probability of the particle toward $\omega \in S^{N-1}$ per unit of time and the computation time, respectively, where $S^{N-1} = \{\omega \in \mathbf{R}^N \mid |\omega| = 1\}$. Then one is led to the master equation

$$\frac{1}{\Delta t}\{q(x, t + \Delta t) - q(x, t)\} = \int_{S^{N-1}} T(x + \omega\Delta x, t; -\omega)q(x + \omega\Delta x, t)d\omega$$

$$- \int_{S^{N-1}} T(x, t; \omega)d\omega \cdot q(x, t) \tag{3.7}$$

and the mean waiting time τ is reformulated via

$$\int_{S^{N-1}} T(x, t; \omega)\,d\omega = \tau^{-1}. \tag{3.8}$$

Here $d\omega$ is an isotropic probability measure on S^{N-1} satisfying

$$\int_{S^{N-1}} d\omega = 1, \quad \int_{S^{N-1}} \omega\,d\omega = 0,$$

$$\int_{S^{N-1}} \omega_i \omega_j\,d\omega = \frac{\delta_{ij}}{N}, \quad \omega = (\omega_i)_{1 \le i \le N}. \tag{3.9}$$

We first assume that the right-hand side on (3.8) is independent of (x, t) and define the diffusion coefficient D by (3.6). If $T(x, t; \omega)$ is a constant denoted by T, it follows from (3.6) and (3.8) that

$$T = \frac{2ND}{(\Delta x)^2}.$$

Then the master equation (3.7) reduces to

$$\frac{1}{\Delta t}\{q(x, t + \Delta t) - q(x, t)\}$$

$$= T \int_{S^{N-1}} q(x + \omega \Delta x, t) - q(x, t)\,d\omega$$

$$= \frac{2ND}{(\Delta x)^2} \int_{S^{N-1}} q(x + \omega \Delta x, t) - q(x, t)\,d\omega. \tag{3.10}$$

Using the elementary lemma below, we obtain

$$q_t = D\Delta q \tag{3.11}$$

as the mean-field limit when $\Delta t \downarrow 0$ and $\Delta x \downarrow 0$ in (3.10).

Lemma 3.1.1 *If $f = f(x)$ is a C^2-function, then*

$$\int_{S^{N-1}} \{f(x + \omega \Delta x) - f(x)\}\,d\omega = \frac{(\Delta x)^2}{2N} \Delta f(x) + o\left((\Delta x)^2\right)$$

as $\Delta x \to 0$.

The general case (3.3) is also consistent with (3.6). First, if $T(x, t; \omega)$ is independent of $\omega \in S^{N-1}$ so denoting it by $T(x, t)$ we have, by (3.8),

$$\tau^{-1} = T(x, t),$$

which by (3.6) yields

$$T(x, t) = \frac{2ND(x, t)}{(\Delta x)^2}.$$

Now the master equation (3.7) reads

$$\frac{1}{\Delta t}\{q(x, t + \Delta t) - q(x, t)\}$$

$$= \int_{S^{N-1}} \{T(x + \omega\Delta x, t)q(x + \omega\Delta x, t) - T(x, t)q(x, t)\}\ d\omega$$

$$= \frac{2N}{(\Delta x)^2} \int_{S^{N-1}} \{D(x + \omega\Delta x, t)q(x + \omega\Delta x, t) - D(x, t)q(x, t)\}\ d\omega,$$

$$(3.12)$$

which reduces to the mean-field equation

$$q_t = \Delta(Dq).$$

The Smoluchowski equation is also derived using a renormalized barrier [54]. Thus we assume that

$$T(x, t; \omega) = cT\left(x + \frac{\Delta x}{2}\omega, t\right)$$

in (3.7) and adjust c so that the mean waiting time τ is uniform:

$$\int_{S^{N-1}} T(x, t; \omega)d\omega = c \int_{S^{N-1}} T\left(x + \frac{\Delta x}{2}\omega, t\right) d\omega = \tau^{-1},$$

which means that

$$T(x, t; \omega) = \frac{\tau^{-1}T\left(x + \frac{\Delta x}{2}\omega, t\right)}{\int_{S^{N-1}} T\left(x + \frac{\Delta x}{2}\omega', t\right) d\omega'} \tag{3.13}$$

(Fig. 3.1).

Then the master equation (3.7), Einstein's formula (3.6), and the transition probability formula (3.13) yield the Smoluchowski equation

$$q_t = D\nabla \cdot (\nabla q - q\nabla \log T) \tag{3.14}$$

with constant diffusion coefficient D.

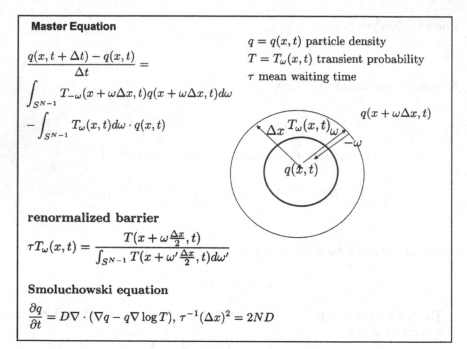

Fig. 3.1   Renormalized barrier

In fact, Eq. (3.7) with (3.13) reads

$$\frac{\tau}{\Delta t}\{q(x, t + \Delta t) - q(x, t)\}$$

$$= \tau \int_{S^{N-1}} \{T(x + \omega\Delta x, t; -\omega)q(x + \omega\Delta x, t) - T(x, t; \omega)q(x, t)\}\, d\omega$$

$$= \int_{S^{N-1}} T\left(x + \frac{\Delta x}{2}\omega, t\right) \cdot \left[\frac{q(x + \omega\Delta x, t)}{\int_{S^{N-1}} T(x + \omega\Delta x + \frac{\Delta x}{2}\omega', t)d\omega'}\right.$$

$$\left. - \frac{q(x, t)}{\int_{S^{N-1}} T(x + \frac{\Delta x}{2}\omega', t)d\omega'}\right] d\omega. \tag{3.15}$$

We have

$$T\left(x + \omega\Delta x + \frac{\Delta x}{2}\omega', t\right) = T(x, t) + (\Delta x)\left(\omega + \frac{\omega'}{2}\right) \cdot \nabla T(x, t)$$

$$+ \frac{1}{2}(\Delta x)^2 \left(\omega + \frac{\omega'}{2}\right) \cdot K(x, t)\left(\omega + \frac{\omega'}{2}\right) + o\left((\Delta x)^2\right),$$

and hence

$$\int_{S^{N-1}} T\left(x + \omega\Delta x + \frac{\Delta x}{2}\omega', t\right)d\omega' = T(x, t) + (\Delta x)\omega \cdot \nabla T(x, t)$$

$$+ \frac{1}{2}(\Delta x)^2 \left(\omega \cdot K(x, t)\omega + \frac{1}{4N}\Delta T(x, t)\right) + o\left((\Delta x)^2\right)$$

for $K = \left(\dfrac{\partial^2 T}{\partial x_i \partial x_j}\right)$. Then it holds that

$$\left\{\int_{S^{N-1}} T\left(x + \omega\Delta x + \frac{\Delta x}{2}\omega', t\right)d\omega'\right\}^{-1}$$

$$= \frac{1}{T(x, t)}\{1 - (\Delta x)\omega \cdot T(x, t)^{-1}\nabla T(x, t)$$

$$- \frac{1}{2}(\Delta x)^2 T(x, t)^{-1}(\omega \cdot K(x, t)\omega + \frac{1}{4N}\Delta T(x, t))$$

$$+ (\Delta x)^2 T(x, t)^{-2}(\omega \cdot \nabla T(x, t))^2\} + o\left((\Delta x)^2\right). \qquad (3.16)$$

Since

$$q(x + \omega\Delta x, t) = q(x, t) + (\Delta x)\omega \cdot \nabla q(x, t) + \frac{1}{2}(\Delta x)^2 \omega \cdot H(x, t)\omega + o((\Delta x)^2)$$

for $H = \left(\dfrac{\partial^2 q}{\partial x_i \partial x_j}\right)$, one has that

$$\frac{q(x + \omega\Delta x, t)}{\int_{S^{N-1}} T(x + \omega\Delta x + \frac{\Delta x}{2}\omega', t)d\omega'}$$

$$= \{q(x, t) + (\Delta x)\omega \cdot \nabla q(x, t) + \frac{1}{2}(\Delta x)^2 \omega \cdot H(x, t)\omega\} \cdot$$

$$\cdot \frac{1}{T(x, t)} \cdot \left[1 - (\Delta x)\omega \cdot T(x, t)^{-1}\nabla T(x, t)\right.$$

$$+ (\Delta x)^2\{-\frac{T(x, t)^{-1}}{2}(\omega \cdot K(x, t)\omega + \frac{1}{4N}\Delta T(x, t))$$

$$\left. + T(x, t)^{-2}(\omega \cdot \nabla T(x, t))^2\}\right] + o\left((\Delta x)^2\right)$$

$$= \frac{1}{T(x, t)}\left\{q(x, t) + (\Delta x)\left[\omega \cdot \nabla q(x, t) - q(x, t)\omega \cdot \nabla \log T(x, t)\right]\right.$$

$$+ (\Delta x)^2\left[\frac{1}{2}\omega \cdot H(x, t)\omega - (\omega \cdot \nabla q(x, t))(\omega \cdot \nabla \log T(x, t))\right.$$

$$+ q(x, t)\left(-\frac{1}{2T(x, t)}(\omega \cdot K(x, t)\omega + \frac{1}{4N}\Delta T(x, t))\right.$$

$$\left.\left.\left. + (\omega \cdot \nabla \log T(x, t))^2\right)\right]\right\} + o\left((\Delta x)^2\right), \qquad (3.17)$$

by (3.16). Similarly,

$$T\left(x + \frac{\Delta x}{2}w', t\right) = T(x, t) + (\Delta x)\frac{w'}{2} \cdot \nabla T(x, t) + \frac{(\Delta x)^2}{8}w' \cdot K(x, t)w' + o((\Delta x)^2)$$

implies

$$\int_{S^{N-1}} T\left(x + \frac{\Delta x}{2}w', t\right) dw' = T(x, t) + \frac{(\Delta x)^2}{8N}\Delta T(x, t) + o\left((\Delta x)^2\right),$$

and therefore,

$$\frac{q(x, t)}{\int_{S^{N-1}} T\left(x + \frac{\Delta x}{2}w', t\right) dw'}$$

$$= \frac{q(x, t)}{T(x, t)} \cdot \left(1 - \frac{(\Delta x)^2}{8N}T(x, t)^{-1}\Delta T(x, t)\right) + o((\Delta x)^2). \qquad (3.18)$$

Now (3.17) and (3.18) yield

$$\frac{q(x + w\Delta x, t)}{\int_{S^{N-1}} T\left(x + w\Delta x + \frac{\Delta x}{2}w', t\right) dw'} - \frac{q(x, t)}{\int_{S^{N-1}} T\left(x + \frac{\Delta x}{2}w', t\right) dw'}$$

$$= \frac{1}{T(x, t)}\left\{(\Delta x)\left[w \cdot \nabla q(x, t) - q(x, t)w \cdot \nabla \log T(x, t)\right]\right.$$

$$+ (\Delta x)^2\left[\frac{1}{2}w \cdot H(x, t)w - (w \cdot \nabla q(x, t))(w \cdot \nabla \log T(x, t))\right.$$

$$\left.+ q(x, t)\left(-\frac{1}{2T(x, t)}w \cdot K(x, t)w + (w \cdot \nabla \log T(x, t))^2\right)\right]\right\}$$

$$+ o\left((\Delta x)^2\right)$$

which implies

$$T\left(x + \frac{\Delta x}{2}w, t\right)$$

$$\cdot \left\{\frac{q(x + w\Delta x, t)}{\int_{S^{N-1}} T\left(x + w\Delta x + \frac{\Delta x}{2}w', t\right) dw'} - \frac{q(x, t)}{\int_{S^{N-1}} T\left(x + \frac{\Delta x}{2}w', t\right) dw'}\right\}$$

$$= \left\{T(x, t) + \frac{\Delta x}{2}w \cdot \nabla T(x, t) + \frac{1}{D}(\Delta x)^2 w \cdot K(x, t)w\right\}$$

$$\cdot \frac{1}{T(x, t)}\left\{(\Delta x)\left[w \cdot \nabla q(x, t) - q(x, t)w \cdot \nabla \log T(x, t)\right]\right.$$

$$+ (\Delta x)^2\left[\frac{1}{2}w \cdot H(x, t)w - (w \cdot \nabla q(x, t))(w \cdot \nabla \log T(x, t))\right.$$

$$+ q(x, t)\left(-\frac{1}{2T(x, t)}\omega \cdot K(x, t)\omega + (\omega \cdot \nabla \log T(x, t))^2\right)\right]\right\}$$
$$+ o((\Delta x)^2).$$

Here, the right-hand side is equal to

$$\left[\omega \cdot \nabla q(x, t) - q(x, t)\omega \cdot \nabla \log T(x, t)\right](\Delta x)$$
$$+ \frac{1}{T(x, t)}\left[\frac{1}{2}(\omega \cdot \nabla T(x, t))(\omega \cdot \nabla q(x, t) - q(x, t)\omega \cdot \nabla \log T(x, t))\right.$$
$$+ T(x, t)(\frac{1}{2}\omega \cdot H(x, t)\omega - (\omega \cdot \nabla q(x, t))(\omega \cdot \nabla \log T(x, t)))$$
$$+ T(x, t)q(x, t)\left(-\frac{1}{2T(x, t)}\omega \cdot K(x, t)\omega + (\omega \cdot \nabla \log T(x, t))^2\right)\right](\Delta x)^2$$
$$+ o((\Delta x)^2).$$

Therefore,

$$\int_{S^{N-1}} T\left(x + \frac{\Delta x}{2}\omega, t\right)$$
$$\cdot \left\{\frac{q(x + \omega\Delta x, t)}{\int_{S^{N-1}} T\left(x + \omega\Delta x + \frac{\Delta x}{2}\omega', t\right) d\omega'} - \frac{q(x, t)}{\int_{S^{N-1}} T\left(x + \frac{\Delta x}{2}\omega', t\right) d\omega'}\right\} d\omega$$
$$= \frac{(\Delta x)^2}{T(x, t)}\int_{S^{N-1}}\left\{\frac{1}{2}(\omega \cdot \nabla T(x, t))\left[\omega \cdot \nabla q(x, t) - q(x, t)\omega \cdot \nabla \log T(x, t)\right]\right.$$
$$+ T(x, t)\left[\frac{1}{2}\omega \cdot H(x, t)\omega - (\omega \cdot \nabla q(x, t))(\omega \cdot \nabla \log T(x, t))\right]$$
$$+ T(x, t)q(x, t)\left[-\frac{1}{2T(x, t)}\omega \cdot K(x, t)\omega + (\omega \cdot \nabla \log T(x, t))^2\right]\right\} d\omega$$
$$+ o((\Delta x)^2).$$

Using the relation

$$\int_{S^N} (a \cdot \omega)(b \cdot \omega)d\omega = \int_{S^{N-1}} \sum_{i,j} a_i b_j \omega_i \omega_j d\omega = \frac{1}{N}\sum a_i b_i = \frac{1}{N}a \cdot b,$$

we end up with

$$\frac{\tau}{\Delta t}\{q(x, t + \Delta t) - q(x, t)\}$$
$$= \frac{(\Delta x)^2}{NT(x, t)}\left\{\frac{1}{2}\left[\nabla T(x, t) \cdot \nabla q(x, t) - q(x, t)\nabla T(x, t) \cdot \nabla \log T(x, t)\right]\right.$$
$$+ T(x, t)\left[\frac{1}{2}\Delta q(x, t) - \nabla q(x, t) \cdot \nabla \log T(x, t)\right]$$

$$+ T(x,t)q(x,t) \cdot \left[-\frac{1}{2T(x,t)} \Delta T(x,t) + |\nabla \log T(x,t)|^2 \right] \right\} + o((\Delta x)^2).$$

Now we apply (3.6) to take the limit $\Delta t \downarrow 0$, $\Delta x \downarrow 0$, and obtain

$$
\begin{aligned}
q_t &= \frac{2D}{T} \left\{ \frac{1}{2} [\nabla T \cdot \nabla q - q \nabla T \cdot \nabla \log T] + T \left[\frac{1}{2} \Delta q - \nabla q \cdot \nabla \log T \right] \right. \\
&\quad \left. + Tq \left[-\frac{1}{2} T^{-1} \Delta T + |\nabla \log T|^2 \right] \right\} \\
&= 2D \left\{ \frac{1}{2} \Delta q + \frac{1}{2} [\nabla \log T \cdot \nabla q - q |\nabla \log T|^2] - \nabla q \cdot \nabla \log T \right. \\
&\quad \left. - \frac{1}{2} q T^{-1} \Delta T + q |\nabla \log T|^2 \right\} \\
&= D \left\{ \Delta q + \nabla \log T \cdot \nabla q + q |\nabla \log T|^2 - 2\nabla q \cdot \nabla \log T - q T^{-1} \Delta T \right\} \\
&= D \nabla \cdot (\nabla q - q \nabla \log T),
\end{aligned}
$$

i.e., the Smoluchowski equation (3.14).

3.2 Random Theory

We have to examine the validity of deterministic formulation, because jump processes are executed randomly in stochastic simulations. Here we take random walk on $\mathbf{R}^N$ where the particle executes successive jumps. Their waiting times are random variables independently and identically distributed. Let $\phi(t), t > 0$, be the probability density for the jump to occur at the time t, and $S(x, y)$ be that for the jump from y to x. Following [100], we assume that these two probabilities are independent and furthermore

$$S(x, y) = S(x - y).$$

Let $Q_k(x, t)$ be the conditional probability that the particle located in $x = 0$ at $t = 0$ reaches x at t after k-steps. The master equation then takes the form

$$Q_k(x, t) = \int_0^t \int_{\mathbf{R}^N} \phi(t - \tau) S(x - y) Q_{k-1}(y, \tau) dy d\tau.$$

It follows that the probability density that the particle reaches x at t is given by

$$
\begin{aligned}
Q(x, t) &= \sum_{k=0}^{\infty} Q_k(x, t) \\
&= Q_0(x, t) + \int_0^t \int_{\mathbf{R}^N} \phi(t - \tau) S(x - y) \sum_{k=0}^{\infty} Q_k(y, \tau) \, dy d\tau
\end{aligned}
$$

$$= Q_0(x,t) + \int_0^t \int_{\mathbf{R}^N} \phi(t-\tau)S(x-y)Q(y,\tau)\,dy d\tau.$$

Since

$$Q_0(x,t) = \delta(x)\delta(t),$$

it holds that

$$Q(x,t) = \delta(x)\delta(t) + \int_0^t \int_{\mathbf{R}^N} \phi(t-\tau)S(x-y)Q(y,\tau)\,dy d\tau.$$

Let $q(x,t)$ be the probability that the particle located in x at $t=0$ reaches the position x at the time t. Then it holds that

$$q(x,t) = \int_0^t \Phi(t,\tau;x)Q(x,\tau)d\tau$$

where $\Phi(t,\tau;x)$ denotes the probability density that the particle reaches x at $\tau < t$ and it does not jump in the remaining time. From the assumption on the jump process we obtain that $\Phi(t,\tau;x) = \Phi(t-\tau)$, with

$$\Phi(t) = \int_t^\infty \phi(s)ds = 1 - \int_0^t \phi(s)ds,$$

which stands for the probability density that the particle does not move in the time interval $(0,t)$. Then it follows that

$$q(x,t) = \int_0^t \Phi(t-\tau)Q(x,\tau)d\tau$$
$$= \int_0^t \Phi(t-\tau)\left\{\delta(x)\delta(\tau) + \int_0^\tau \int_{\mathbf{R}^N} \phi(\tau-s)S(x-y)Q(y,s)\,dyds\right\}d\tau$$
$$= \Phi(t)\delta(x) + \int_0^t \int_{\mathbf{R}^N}\left\{\int_s^t \Phi(t-\tau)\phi(\tau-s)d\tau\right\}S(x-y)Q(y,s)dyds$$
$$= \Phi(t)\delta(x) + \int_0^t \int_{\mathbf{R}^N}\left\{\int_s^t \phi(t-\tau)\Phi(\tau-s)d\tau\right\}S(x-y)Q(y,s)dyds$$
$$= \Phi(t)\delta(x) + \int_0^t \int_{\mathbf{R}^N} \phi(t-\tau)S(x-y)q(y,\tau)dyd\tau. \tag{3.19}$$

The requirements for these probability densities are

$$\int_0^\infty \phi(t)dt = 1, \quad \int_{\mathbf{R}^N} S(x)dx = 1.$$

Using the Laplace and Fourier transforms of $f(t)$ and $g(x)$,

$$\mathcal{L}\{f(t); s\} = \tilde{f}(s) = \int_0^\infty e^{-st} f(t)dt$$

and

$$\mathcal{F}\{g(x); k\} = \hat{g}(k) = \int_{\mathbf{R}^N} e^{ik \cdot x} g(x)dx,$$

respectively, we obtain

$$\hat{\tilde{q}}(k, s) = \frac{1 - \tilde{\phi}(s)}{s} + \tilde{\phi}(s)\hat{S}(k)\hat{\tilde{q}}(k, s). \qquad (3.20)$$

Let

$$\tilde{H}(s) = \frac{\tilde{\phi}(s)}{1 - \tilde{\phi}(s)}.$$

Then, (3.20) reads

$$\frac{\hat{\tilde{q}}(k, s)}{1 - \tilde{\phi}(s)} - \frac{1}{s} = \tilde{H}(s)\hat{S}(k)\hat{\tilde{q}}(k, s).$$

Subtracting

$$\frac{\hat{\tilde{q}}(k, s)\tilde{\phi}(s)}{1 - \tilde{\phi}(s)}$$

from both sides, we obtain

$$\hat{\tilde{q}}(k, s) - \frac{1}{s} = (\hat{S}(k) - 1)\tilde{H}(s)\hat{\tilde{q}}(k, s). \qquad (3.21)$$

Inverting (3.21) back to the space-time domain yields

$$q(x, t) - q(x, 0)$$
$$= \int_0^t H(t - \tau)\left\{-q(x, \tau) + \int_{\mathbf{R}^N} S(x - y)q(y, \tau)dy\right\} d\tau. \qquad (3.22)$$

When the waiting time probability density is Poissonian,

$$\phi(t) = \frac{e^{-t/\lambda}}{\lambda}, \quad \lambda > 0,$$

we have

$$\tilde{\phi}(s) = \int_0^\infty e^{-st} \cdot \frac{e^{-t/\lambda}}{\lambda} dt = \frac{1}{1 + \lambda s}.$$

Then

$$\tilde{H}(s) = \frac{\tilde{\phi}(s)}{1 - \tilde{\phi}(s)} = \frac{1}{\lambda s},$$

and hence $H(t) = 1/\lambda$. Equation (3.22) is now reduced to

$$q(x, t) - q(x, 0) = \frac{1}{\lambda} \int_0^t \left\{ -q(x, \tau) + \int_{\mathbf{R}^N} S(x - y)q(y, \tau)dy \right\} d\tau$$

$$= \frac{1}{\lambda} \int_0^t \int_{\mathbf{R}^N} S(x - y)[q(y, t) - q(x, t)]dy \, d\tau,$$

or

$$q_t(x, t) = \frac{1}{\lambda} \int_{\mathbf{R}^N} S(x - y)[q(y, t) - q(x, t)] \, dy. \tag{3.23}$$

If

$$S(-x) = S(x)$$

$$\int_{\mathbf{R}^N} x_i x_j S(x)dx = \frac{(\Delta x)^2}{N}\delta_{ij}, \quad 1 \le i, j \le N, \tag{3.24}$$

then we obtain

$$\frac{1}{\lambda} \int_{\mathbf{R}^N} S(x - y)(q(y, t) - q(x, t))dy = \frac{1}{\lambda} \int_{\mathbf{R}^N} S(y - x) \times$$

$$\times \left\{ (y - x) \cdot \nabla q(x, t) + \frac{1}{2}[\nabla^2 q(x, t)](y - x) \cdot (y - x) + o(|y - x|^2) \right\} dy$$

$$= \frac{(\Delta x)^2}{2N\lambda}(\Delta q + o(1)) = D\Delta q + o(1)$$

similarly to Lemma 3.1.1. Now from (3.24) and (3.6) with $\tau = \lambda$ it follows that

$$q_t = D\Delta q$$

as $\Delta x \to 0$.

The master equation (3.22) derived from the random theory can model anomalous diffusions. Subdiffusion is one of them, characterized by the power-law pattern of the mean squared displacement at time t,

$$\langle x(t)^2 \rangle \sim t^\alpha, \quad 0 < \alpha < 1.$$

This profile is different from that of a particle moving in normal diffusion, which follows the pattern

$$\langle x(t)^2 \rangle \sim t \tag{3.25}$$

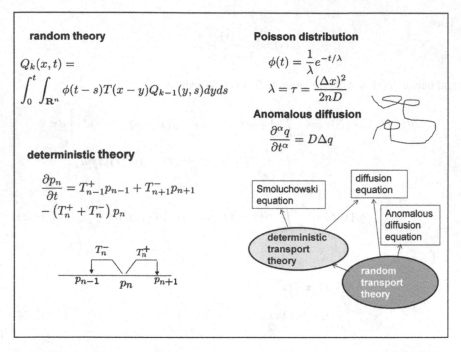

Fig. 3.2 Mean field approximations—deterministic and random

ensured by (3.6). Many processes can no longer be described by the normal diffusion process (3.25), for example, migration of contaminants in groundwater systems [8], dispersive transport of ions in column experiments [41], diffusion of water in sand [77], dispersion in a heterogeneous aquifer [1], and transport of contaminants in geological formations [7]. The model process observed in these cases is subdiffusion, and an appropriate reaction-subdiffusion equation has been derived. The model of [113] has a memory effect in both the reaction and diffusion terms, while that of [121] has it only in the diffusion term (Fig. 3.2).

3.3   Stochastic Simulations

Molecular events in biological cells occur in local subregions where the molecules tend to be small in number. In this situation, the concentration is not a good measure. Hence deterministic simulation using ordinary differential equations is not an appropriate tool and stochastic simulations should be run to simulate the local environment. In addition, stochastic simulation should converge to a macroscopic law of mass action when the number of molecules is sufficiently large.

Following [53], we assume a reaction radius centered at a reacting mole-cule in 3D space, within which a reaction occurs with the rate P_r. Using the average number of molecules within the reaction radius and the number of jumps for one molecule within a short period of time, we show a formula calculating P_r from the reaction rate k. Each reaction proceeds using this P_r and number of molecules within each reaction radius. Thus, a reaction between the two molecules takes place in a fully local environment, that is, each reaction depends only on the occurrence of a collision of RW (randomly walking) molecules. This method allows multiple collisions within a computation time, and the earliest collision is selected for a reaction.

In the stochastic simulation performed in [53], no mesh division of the $3D$ space is used, and molecules can jump in any direction. Thus, each molecule changes its coordinates by jumping from x_t to $x_{t+\tau}$ with a waiting time τ. If two reacting molecules meet within the reaction radius, a reaction takes place at a rate P_r ($0 < P_r < 1$). This poses the problem of finding a formula for calculating P_r from k. The stochastic simulation of the first-order reaction is realized by random transition with average P_r, which is calculated from the exponential function defined by k as in [6]. These processes fit the master equation in Sect. 3.1. With the constant mean waiting time and jump length, then we get a reaction-diffusion equation with non-local term in a mean field limit.

Taking the fundamental process

$$A + B \rightarrow P \ (k), \qquad (3.26)$$

we adopt the ansatz that a chemical reaction occurs if and only if two A, B particles lie within in the distance R from one another. We call this R the *reaction radius* and denote the reaction probability at each collision of particles A, B by P_r (Fig. 3.3).

Given an A-particle, the number of B-particles within a distance equal to the reaction radius from it is $n_B = [B]N_a v$, where $[B]$ is the concentration of B-particles, N_a is the Avogadro number, and $v = \omega_N R^N$, with ω_N denoting the volume of the N-dimensional unit ball. If Q_A and n_{jA} denote the number of A-particles in the vessel and the number of jumps of each A-particle per unit of time, respectively, then

$$\frac{dQ_{A,A\rightarrow B}}{dt} = -P_r Q_A n_{jA} n_B$$

stands for the rate of change of the number of A-particles reacting to B-particles as a result of the jumps of A-particles. Letting V be the volume of the vessel and $[A]_{A\rightarrow B} = \frac{Q_{A,A\rightarrow B}}{N_a V}$, it holds that

$$\frac{d[A]_{A\rightarrow B}}{dt} = -P_r[A]n_{jA}n_B = -P_r N_a v n_{jA}[A][B],$$

since $Q_A = [A]N_a V$. The relation

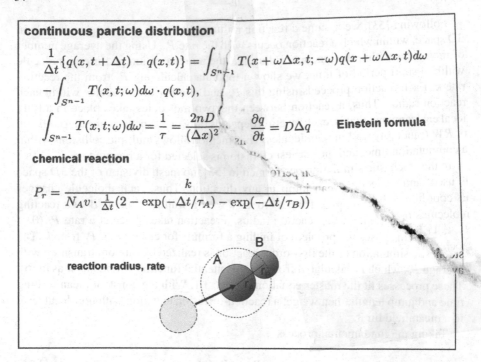

continuous particle distribution

$$\frac{1}{\Delta t}\{q(x,t+\Delta t) - q(x,t)\} = \int_{S^{n-1}} T(x+\omega\Delta x, t; -\omega)q(x+\omega\Delta x, t)d\omega$$

$$-\int_{S^{n-1}} T(x,t;\omega)d\omega \cdot q(x,t),$$

$$\int_{S^{n-1}} T(x,t;\omega)d\omega = \frac{1}{\tau} = \frac{2nD}{(\Delta x)^2} \qquad \frac{\partial q}{\partial t} = D\Delta q \qquad \textbf{Einstein formula}$$

chemical reaction

$$P_r = \frac{k}{N_A v \cdot \frac{1}{\Delta t}(2 - \exp(-\Delta t/\tau_A) - \exp(-\Delta t/\tau_B))}$$

reaction radius, rate

Fig. 3.3   Mean field approximation—chemical reaction

$$\frac{d[A]_{B \to A}}{dt} = -P_r N_a v n_{jB}[A][B]$$

governing the rate of change of the concentration of A-particles as a result of the jump of B-particles is obtained similarly. Hence it holds that

$$\frac{d[A]}{dt} = -P_r N_a v (n_{jA} + n_{jB})[A][B], \tag{3.27}$$

where n_{jB} the number of jumps of a B-particle per unit time. Comparing (3.27) with the mass reaction equation

$$\frac{d[A]}{dt} = -k[A][B], \tag{3.28}$$

we obtain

$$k = P_r N_a v (n_{jA} + n_{jB}). \tag{3.29}$$

This formula determines the chemical reaction probability which does not depend on the position of the particle so that no effects from the boundary of the vessel are involved.

Here the Poisson process is used in the jump probability [53]. Since the distribution function of this process is given by $p(t) = (1/\tau)\exp(-t/\tau)$, the number of particles jumping per unit time is averaged by

$$n_j = \frac{1}{\Delta t} \int_0^{\Delta t} p(t)dt = \frac{1}{\Delta t}\{1 - \exp(-\Delta t/\tau)\}$$

during the computation time Δt, which implies that

$$n_{jA} = \frac{1}{\Delta t}\{1 - \exp(-\Delta t/\tau_A)\},$$

$$n_{jB} = \frac{1}{\Delta t}\{1 - \exp(-\Delta t/\tau_B)\}. \tag{3.30}$$

We obtain

$$P_r = \frac{k}{N_a v \cdot \frac{1}{\Delta t}\{2 - e^{-\Delta t/\tau_A} - e^{-\Delta t/\tau_B}\}}, \tag{3.31}$$

by (3.29). Hence we require

$$\lim_{\Delta t \downarrow 0} P_r = \frac{k}{N_a v \left(\frac{1}{\tau_A} + \frac{1}{\tau_B}\right)} \in (0, 1).$$

Adopting the notion of reaction radius in the master equation, we use the existence probabilities of A and B-particles, denoted by $q_A = q_A(x, t)$ and $q_B = q_B(x, t)$, respectively. These quantities are dimensionless and it holds that $0 \le q_A, q_B \le 1$. Hence we assume that

$$q_A = [A]/[A]_*, \qquad q_B = [B]/[B]_*$$

in the spatially homogeneous solution, where $[A]_*$ and $[B]_*$ denote the saturated concentrations of the A- and B - particles, respectively. The mass reaction (3.28) then reads

$$\frac{dq_A}{dt} = -k_A q_A q_B, \qquad \frac{dq_B}{dt} = -k_B q_A q_B, \tag{3.32}$$

where $k_A = k/[B]_*$ and $k_B = k/[A]_*$.

From the law of chemical reaction using the concentration radius $R > 0$, the master equation is now stated as

$$\frac{1}{\Delta t}\{q_A(x, t + \Delta t) - q_A(x, t)\}$$

$$= \int_{S^{N-1}} (T_A(x + \omega \Delta x, t; -\omega)q_A(x + \omega \Delta x, t) - T_A(x, t; \omega)q_A(x, t)) \, d\omega$$

$$- \frac{k^{B \to A}}{v} \int_{B(x,R) \cap \Omega} q_B(y,t) dy \cdot q_A(x,t) \tag{3.33}$$

where $k^{B \to A}$ denotes the relative reaction rate of A-particle caused by the collision of B-particle. This value $k^{B \to A}$ is associated with $[A]_*$ and $[B]_*$, which is determined later.

In (3.33) we take $v = |B(\cdot, R)| = \omega_N R^N$ for the normalization of the total number of B-particles inside $B(x, R) \cap \Omega$. Similarly, we obtain

$$\frac{1}{\Delta t} \{q_B(x, t + \Delta t) - q_B(x, t)\}$$

$$= \int_{S^{N-1}} (T_B(x + \omega \Delta x, t; -\omega) q_B(x + \omega \Delta x, t) - T_B(x, t; \omega) q_B(x, t)) \, d\omega$$

$$- \frac{k^{A \to B}}{v} \int_{B(x,R) \cap \Omega} q_A(y, t) dy \cdot q_B(x, t).$$

The limit system

$$\frac{\partial q_A}{\partial t} = D_A \Delta q_A - \frac{k^{B \to A}}{v} \int_{B(\cdot, R) \cap \Omega} q_B dy \cdot q_A,$$

$$\frac{\partial q_B}{\partial t} = D_B \Delta q_B - \frac{k^{A \to B}}{v} \int_{B(\cdot, R) \cap \Omega} q_A dy \cdot q_B, \tag{3.34}$$

arises with constant transition probabilities, and then (3.34) is consistent to (3.32) if and only if

$$k^{B \to A} = k_A = k[B]_*, \qquad k^{A \to B} = k_B = k[A]_*.$$

Now we define the proper computation time Δt by

$$k^{A \to B} + k^{B \to A} = \frac{P_r}{\Delta t},$$

which implies that

$$[A]_* + [B]_* = \frac{1}{N_a v \{2 - e^{-\Delta t / \tau_A} - e^{-\Delta t / \tau_B}\}}$$

and hence

$$N_a v \cdot \Delta t \sim \frac{1}{(\frac{1}{\tau_A} + \frac{1}{\tau_B})([A]_* + [B]_*)}, \qquad 0 < \Delta t \ll 1.$$

If we assume the common jump length Δx for both A and B particles, then, by (3.6),

$$\frac{(\Delta x)^2}{\Delta t} \sim N_a v \cdot 2N(D_A + D_B)([A]_* + [B]_*). \tag{3.35}$$

Equation (3.35) shows how v, or the reaction radius $R > 0$, is associated with the discretization ratio $\gamma = (\Delta x)^2/(\Delta t)$.

3.4 Segment Model

Here we apply the deterministic theory in Sect. 3.1 to actin polymerization. For simplicity we regard each F-actin as a segment.

First, we construct a model without polymerization. Let

$$q = q(x, t, e)$$

be the existence probability at the position of the center $x \in \mathbf{R}^N$, the time $t > 0$, and in the direction of axis denoted by $e \in S^{N-1}$. This axis rotates randomly so that the transition probability T per unit time is denoted by

$$T = T(x, t, \omega, A),$$

where $\omega \in S^{N-1}$ and $A^{-1} \in O(N)$ stand for the jump direction of the center of the segment and the rotation of the axis, respectively. The master equation, therefore, reads

$$\frac{1}{\Delta t}\{q(x, t + \Delta t, e) - q(x, t, e)\}$$

$$= \iint_{S^{N-1} \times O(N)} T(x + \omega \Delta x, t, -\omega, A)q(x + \omega \Delta x, t, A^{-1}e)d\omega dA$$

$$- \iint_{S^{N-1} \times O(N)} T(x, t, \omega, A)q(x, t, e)d\omega dA \qquad (3.36)$$

with the probability measures $d\omega$ and dA on S^{N-1} and $O(N)$, respectively:

$$\int_{S^{N-1}} d\omega = 1, \qquad \int_{O(N)} dA = 1.$$

Then Einstein's formula yields

$$\iint_{S^{N-1} \times O(N)} T(x, t, \omega, A)d\omega dA = \frac{1}{\tau} = \frac{2ND}{(\Delta x)^2}.$$

In the case where $T(x, t, \omega, A)$ is a constant denoted by $T > 0$, we have

$$T = \frac{2ND}{(\Delta x)^2} \qquad (3.37)$$

and hence the master equation (3.36) becomes

$$\frac{1}{\Delta t}\{q(x, t + \Delta t, e) - q(x, t, e)\}$$

$$= \frac{2ND}{(\Delta x)^2} \iint_{S^{N-1} \times O(N)} \left[q(x + w\Delta x, t, A^{-1}e) - q(x, t, e) \right] dw dA.$$

We assume the invariance of dA under the transformations $A \mapsto A^{-1}$ and $A \mapsto BA$, $A \mapsto AB$, and also that of dw under the transformation $w \mapsto Bw$, where $B \in O(N)$ is arbitrary. Given $e', e \in S^{N-1}$, we take $B \in O(N)$ such that $e' = Be$ and then,

$$\int_{O(N)} f(Ae)dA = \int_{O(N)} f(Ae')dA$$

for all $f \in C(S^{N-1})$. We have also

$$\int_{S^{N-1}} f(Aw)dw = \int_{S^{N-1}} f(A'w)dw$$

for arbitrary $A, A' \in O(N)$, and hence

$$\int_{O(N)} f(Ae)dA = \iint_{S^{N-1} \times O(N)} f(Aw)dw dA = \int_{S^{N-1}} f(w)dw, \qquad (3.38)$$

by averaging. It follows that

$$\int_{O(N)} q(x, t, A^{-1}e)dA = \int_{O(N)} q(x, t, Ae)dA$$

$$= \int_{S^{N-1}} q(x, t, e')de' \equiv \overline{q}(x, t) \qquad (3.39)$$

and hence

$$\frac{1}{\Delta t}\{q(x, t + \Delta t, e) - q(x, t, e)\}$$

$$= \frac{2ND}{(\Delta x)^2} \int_{S^{N-1}} \{\overline{q}(x + w\Delta x, t) - q(x, t, e)\} \, dw. \qquad (3.40)$$

Although the probability density $q(x, t, e)$ cannot take the mean field limit in (3.40), the averaged probability $\overline{q}(x, t)$ defined in (3.39) satisfies

$$\frac{1}{\Delta t}\{\overline{q}(x, t + \Delta t) - \overline{q}(x, t)\}$$

$$= \frac{2ND}{(\Delta x)^2} \int_{S^{N-1}} \overline{q}(x + w\Delta x, t) - \overline{q}(x, t)dw. \qquad (3.41)$$

Under the isotropy assumption (3.9), Eq. (3.41) yields the mean field limit

$$\frac{\partial \overline{q}}{\partial t} = D \Delta \overline{q}, \tag{3.42}$$

thanks to Lemma 3.1.1.

Relations (3.40)–(3.41) now imply

$$\frac{1}{\Delta t}\{q(x, t + \Delta t, e) - q(x, t, e)\}$$

$$= \frac{2ND}{(\Delta x)^2} \int_{S^{N-1}} \{\overline{q}(x + \omega \Delta x, t) - \overline{q}(x, t)\}\, d\omega$$

$$+ \frac{2ND}{(\Delta x)^2} \int_{S^{N-1}} \{\overline{q}(x, t) - q(x, t, e)\}\, d\omega$$

$$= \frac{1}{\Delta t}\{\overline{q}(x, t + \Delta t) - \overline{q}(x, t)\} + \frac{2ND}{(\Delta x)^2}\{\overline{q}(x, t) - q(x, t, e)\}.$$

Denoting

$$r(x, t, e) = q(x, t, e) - \overline{q}(x, t),$$

we have

$$\int_{S^{N-1}} r(x, t, e)\, de = 0$$

$$\frac{1}{\Delta t}\{r(x, t + \Delta t, e) - r(x, t, e)\} = -\frac{2ND}{(\Delta x)^2} r(x, t, e) = -\frac{1}{\tau} r(x, t, e),$$

and hence

$$r(x, t_n, e) = (1 - \tau^{-1}\Delta t)^n r(x, 0, e), \quad t_n = n\Delta t. \tag{3.43}$$

If $0 < \Delta t \ll 1$ and $\tau \sim 1$, Eq. (3.43) for $r = r(x, t, e)$ is approximated by

$$\frac{\partial r}{\partial t} = -\tau^{-1} r$$

which guarantees, for example, that

$$\int_{S^{N-1}} r(x, t, e)^2\, de \sim \int_{S^{N-1}} r(x, 0, e)^2\, de \cdot \exp(-2\tau^{-1}t), \quad t \uparrow +\infty.$$

This homogenization property is reasonable, regarding (3.38).

We now turn to the polymerization of F-actins between G-actins, the latter assumed to be in sufficient quantity and continuously distributed. This process can be represented by

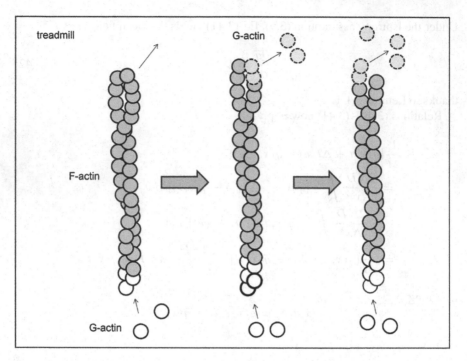

Fig. 3.4 Actin treadmill

$$G + F_\ell \to F_{\ell+1} \ (k_+^\ell), \qquad F_{\ell+1} \to G + F_\ell \ (k_-^\ell)$$

where F_ℓ denotes a chain of F-actin composed of ℓ-G-actins. Adhesion and abrasion occur at the bottom and the top of a F-actin chain, respectively. Let the reaction radii of these processes be R and r, respectively, and let $Q = Q(x, t)$ be the existence probability of G-actins, assumed to be continuous in space and time (Fig. 3.4). Then (3.36) may be modified as

$$\frac{1}{\Delta t}\{q_\ell(x, t + \Delta t, e) - q_\ell(x, t, e)\} = I + II + III + IV - V, \qquad (3.44)$$

with

$$I = \iint_{S^{N-1} \times O(N)} \Big[T_{\ell-1}(x + \omega\Delta x, t, -\omega, A) \times$$

$$\times \frac{P_{r-}^{G \to F_{\ell-1}}}{|B_R|} \int_{B_R(x+\omega\Delta x - \frac{\ell-1}{2}mA^{-1}e)\cap\Omega} Q(y, t)dy \times$$

$$\times \left\{ 1 - \frac{P_{r+}^{G \to F_{\ell-1}}}{|B_r|} \int_{B_r(x+\omega\Delta x + \frac{\ell-1}{2}mA^{-1}e)\cap\Omega} Q(y, t)dy \right\} \times$$

$$\times q_{\ell-1}(x + \omega\Delta x, t, A^{-1}e)\Big] \, d\omega \, dA,$$

$$II = \iint_{S^{N-1}\times O(N)} \Big[T_{\ell+1}(x + \omega\Delta x, t, -\omega, A) \times$$

$$\times \frac{P_{r+}^{G\to F_{\ell+1}}}{|B_r|} \int_{B_r(x+\omega\Delta x + \frac{\ell+1}{2}mA^{-1}e)\cap\Omega} Q(y,t)dy \times$$

$$\times \Big\{ 1 - \frac{P_{r-}^{G\to F_{\ell+1}}}{|B_R|} \int_{B_R(x+\omega\Delta x + \frac{\ell+1}{2}mA^{-1}e)\cap\Omega} Q(y,t)dy \Big\} \times$$

$$\times q_{\ell+1}(x + \omega\Delta x, t, A^{-1}e)\Big] \, d\omega \, dA,$$

$$III = \iint_{S^{N-1}\times O(N)} \Big[T_{\ell}(x + \omega\Delta x, t, -\omega, A) \times$$

$$\times \Big\{ 1 - \frac{P_{r+}^{G\to F_{\ell}}}{|B_r|} \int_{B_r(x+\omega\Delta x + \frac{\ell}{2}mA^{-1}e)\cap\Omega} Q(y,t)dy \Big\} \times$$

$$\times \Big\{ 1 - \frac{P_{r-}^{G\to F_{\ell}}}{|B_R|} \int_{B_R(x+\omega\Delta x - \frac{\ell}{2}mA^{-1}e)\cap\Omega} Q(y,t)dy \Big\} \times$$

$$\times q_{\ell}(x + \omega\Delta x, t, A^{-1}e)\Big] \, d\omega \, dA,$$

$$IV = \iint_{S^{N-1}\times O(N)} \Big[T_{\ell}(x + \omega\Delta x, t, -\omega, A) \times$$

$$\times \frac{P_{r+}^{G\to F_{\ell}}}{|B_r|} \int_{B_r(x+\omega\Delta x + \frac{\ell}{2}mA^{-1}e)\cap\Omega} Q(y,t)dy \times$$

$$\times \frac{P_{r-}^{G\to F_{\ell}}}{|B_R|} \int_{B_R(x+\omega\Delta x - \frac{\ell}{2}mA^{-1}e)\cap\Omega} Q(y,t)dy \times$$

$$\times q_{\ell}(x + \omega\Delta x, t, A^{-1}e)\Big] \, d\omega \, dA,$$

and

$$V = \iint_{S^{N-1}\times O(N)} T_{\ell}(x, t, \omega, A)q_{\ell}(x, t, e) \, d\omega \, dA.$$

Here m, $T_{\ell} = T_{\ell}(x, t, \omega, e)$, and $P_{r-(+)}^{F_{\ell}\to G}$ are the unit length of the G-actin, the transition probability of F_{ℓ}, and the reaction rate of adhesion (abrasion), respectively, that is,

$$P_{r\mp}^{G\to F_\ell} = \frac{k_\pm^\ell}{[F_\ell]_*},$$

where $[F_\ell]_*$ denotes the saturated concentration of F_ℓ. We also make the convention $q_{-1} = 0$.

Then Einstein's formula is described by

$$\iint_{S^{N-1}\times O(N)} T_\ell(x,t,\omega,A)\,d\omega\,dA = \frac{2ND_\ell}{(\Delta x)^2},$$

where D_ℓ denotes the diffusion coefficient of F_ℓ. The computation time Δt and the jump length are assumed to be independ of ℓ for simplicity. We also assume the independence of $P_{r\pm}^{G\to F_\ell}$ on ℓ, putting

$$\mu_\pm = P_{r\pm}^{G\to F_\ell}.$$

If $T = T_\ell(x,t,\omega,A)$ is a constant, then D_ℓ must be independent of ℓ, and we denote it by D. Then (3.37) holds. We also take the limits r, $R \downarrow 0$, and finally, put $Q \equiv 1$ for simplicity. With these simplifications we obtain from (3.44)

$$\frac{1}{\Delta t}\{q_\ell(x,t+\Delta t,e) - q_\ell(x,t,e)\}$$
$$= \frac{2ND}{(\Delta x)^2}\iint_{S^{N-1}\times O(N)}\Big[\mu_-(1-\mu_+)q_{\ell-1}(x+\omega\Delta x,t,A^{-1}e)$$
$$+ \mu_+(1-\mu_-)q_{\ell+1}(x+\omega\Delta x,t,A^{-1}e)$$
$$+ \big[(1-\mu_-)(1-\mu_+)+\mu_+\mu_-\big]q_\ell(x+\omega\Delta t,t,A^{-1}e) - q_\ell(x,t,e)\Big]\,d\omega\,dA.$$

This further reduces to

$$\frac{1}{\Delta t}\{q_\ell(x,t+\Delta t,e) - q_\ell(x,t,e)\}$$
$$= \frac{2ND}{(\Delta x)^2}\int_{S^{N-1}}\Big[\mu_-(1-\mu_+)\overline{q_{\ell-1}}(x+\omega\Delta x,t)$$
$$+ \mu_+(1-\mu_-)\overline{q_{\ell+1}}(x+\omega\Delta x,t)$$
$$+ \big[(1-\mu_-)(1-\mu_+)+\mu_+\mu_-\big]\overline{q_\ell}(x+\omega\Delta t,t)$$
$$- \overline{q_\ell}(x,t)\Big]\,d\omega, \tag{3.45}$$

by (3.38), where

$$\overline{q_\ell}(x,t) = \int_{S^{N-1}} q_\ell(x,t,e)\,de.$$

Equation (3.45) yields

$$\frac{1}{\Delta t}\{\overline{q_\ell}(x, t + \Delta t) - \overline{q_\ell}(x, t)\} = \frac{2ND}{(\Delta x)^2} \int_{S^{N-1}} \Big[\mu_-(1 - \mu_+)\overline{q_{\ell-1}}(x + w\Delta x, t)$$
$$+ \mu_+(1 - \mu_-)\overline{q_{\ell+1}}(x + w\Delta x, t)$$
$$+ \big[(1 - \mu_-)(1 - \mu_+) + \mu_+\mu_-\big]\overline{q_\ell}(x + w\Delta t, t) - \overline{q_\ell}(x, t)\Big] dw, \qquad (3.46)$$

and hence

$$\frac{1}{\Delta t}\{\overline{q}(x, t + \Delta t) - \overline{q}(x, t)\}$$
$$= \frac{2ND}{(\Delta x)^2} \int_{S^{N-1}} \Big[\mu_-(1 - \mu_+)\overline{q}(x + w\Delta x, t) + \mu_+(1 - \mu_-)\overline{q}(x + w\Delta x, t)$$
$$+ \big[(1 - \mu_-)(1 - \mu_+) + \mu_+\mu_-\big]\overline{q}(x + w\Delta t, t) - \overline{q}(x, t)\Big] dw$$
$$= \frac{2ND}{(\Delta x)^2} \int_{S^{N-1}} \Big[\overline{q}(x + w\Delta x, t) - \overline{q}(x, t)\Big] dw \qquad (3.47)$$

for

$$\overline{q}(x, t) = \sum_{\ell=0}^{\infty} \overline{q_\ell}(x, t).$$

Thus we arrive at the mean-field limit

$$\frac{\partial \overline{q}}{\partial t} = D\Delta\overline{q}$$

for this $\overline{q} = \overline{q}(x, t)$.

Chapter 4
Mathematical Analysis

Here we carry out a mathematical analysis using thermodynamical structures. We study Smoluchowski-ODE systems with negative chemotaxis, parabolic systems with non-local term, Hamiltonian formalism for the Lotka-Volterra and Gierer-Meinhardt systems, where the spatially homogeneous part controls the asymptotic behavior of the solution. A dual vairational structure and scaling invariance yield the quantized blowup mechanism to the Smoluchowski-Poisson equation in two-space dimension with the underlying material hierarchy and particle-field duality [122, 123]. In many cases, mesoscopic models are subject to variational structures and scaling invariance. Then we can provide rigorous proof for two important results concerning living organisms: self-organization and homeostasis.

4.1 Negative Chemotaxis

To approach (2.28)–(2.36) we regard (2.28) as a reduction of the full-system of chemotaxis, e.g.,

$$u_t = \nabla \cdot (\nabla u - u\nabla v),$$

$$\tau v_t - \Delta v = u - \frac{1}{|\Omega|} \int_\Omega u \, dx \quad \text{in } \Omega \times (0, T),$$

$$\frac{\partial u}{\partial \nu} - u\frac{\partial v}{\partial \nu} = \frac{\partial v}{\partial \nu} = 0 \quad \text{on } \partial\Omega \times (0, T),$$

$$\int_\Omega v \, dx = 0. \tag{4.1}$$

© Springer Nature Singapore Pte Ltd. 2017
T. Suzuki, *Mathematical Methods for Cancer Evolution*,
Lecture Notes on Mathematical Modelling in the Life Sciences,
DOI 10.1007/978-981-10-3671-2_4

The Smoluchowski-Poisson equation (2.51) derived in Sect. 2.4 is the case $\tau = 0$ of (2.28), while a relative of (2.28) will arise when $\varphi(v) = v$ if the diffusion $-\Delta v$ of v is ignored in the second equation of (2.28), that is,

$$u_t = \nabla \cdot (\nabla u - u \nabla v),$$

$$\tau v_t = u - \frac{1}{|\Omega|} \int_\Omega u \, dx \quad \text{in } \Omega \times (0, T),$$

$$\frac{\partial u}{\partial \nu} - u \frac{\partial v}{\partial \nu} = 0 \qquad \text{on } \partial\Omega \times (0, T).$$

The lack of elliptic regularity in the second equation of the above system, however, makes the v-component unstable even if the interaction is self-repulsive, i.e., if the first equation is replaced by

$$u_t = \nabla \cdot (\nabla u + u \nabla v).$$

The case $\varphi(v) = -v$ of (2.28) is studied by the energy method in [106], where the existence of global-in-time solutions is obtained for the case of one space dimension. In [129], this argument is combined with several estimates derived from the comparison principle for $\varphi(v)$ satisfying (2.36).

Local-in-time well-posedness of (2.28) with (2.29) is not restricted to the case of one space dimension. We have a classical solution uniquely, assuming

$$q_0, v_0 \in C^{2+\alpha}(\overline{\Omega}), \quad \frac{\partial q_0}{\partial \nu} = \frac{\partial v_0}{\partial \nu} = 0 \text{ on } \partial\Omega, \quad q_0 \not\equiv 0 \text{ in } \Omega.$$

Here we show the following theorem.

Theorem 4.1 ([129]) *If $N = 1$, the solution $q = q(x, t)$ to (2.28)–(2.36) is global in time.*

From the construction of a local-in-time solution, the existence time of the solution is estimated from below by the Schauder norm of q. Hence we have only to prove that this norm is bounded in $Q_T = \Omega \times (0, T)$ as far as the solution exists, up to $t = T > 0$. Henceforth, C_i ($i = 1, 2, \ldots, 37$) denote positive constants. First, the embedding (2.39) is described by the following inequality.

Lemma 4.1.1 (Remark 2.1 of [76], p. 74) *If $N = 1$, it holds that*

$$\|p\|_{L^4(0,T;L^\infty(\Omega))} \le C_1 \|p\|_{L^\infty(0,T;L^2(\Omega))}^{1/2} \|\nabla p\|_{L^2(0,T;L^2(\Omega))}^{1/2} \tag{4.2}$$

for $p = p(x) \in L^2(0, T; H^1(\Omega)) \cap L^\infty(0, T; L^2(\Omega))$ satisfying

$$\int_\Omega p = 0.$$

Now we use the following fact.

Lemma 4.1.2 *The function*

$$L = \int_\Omega \left[q(\log q - 1) - \frac{1}{2}\varphi'(v)|\nabla v|^2 \right] dx.$$

is a Lyapunov function for (2.28).

Proof We have

$$\frac{d}{dt} \int_\Omega q(\log q - 1)\, dx = \int_\Omega q_t \log q\, dx = -\int_\Omega q^{-1}|\nabla q|^2\, dx + \int_\Omega \nabla q \cdot \nabla \varphi(v)\, dx$$
$$= I + II$$

and

$$II = \int_\Omega \varphi'(v)\nabla v \cdot \nabla v_t\, dx = \frac{1}{2}\frac{d}{dt}\int_\Omega \varphi'(v)|\nabla v|^2\, dx - \frac{1}{2}\int_\Omega \varphi''(v)v_t|\nabla v|^2\, dx$$
$$= \frac{1}{2}\frac{d}{dt}\int_\Omega \varphi'(v)|\nabla v|^2\, dx - \frac{1}{2}\int_\Omega \varphi''(v)q|\nabla v|^2\, dx,$$

by (2.28). Then it holds that

$$\frac{dL}{dt} = \frac{d}{dt}\int_\Omega \left[q(\log q - 1) - \frac{1}{2}\varphi'(v)|\nabla v|^2\right] dx$$
$$= -\int_\Omega \left[q^{-1}|\nabla q|^2 + \frac{1}{2}\varphi''(v)q|\nabla v|^2\right] dx \le 0, \qquad (4.3)$$

because $\varphi'' \ge 0$. □

Since the function $s(\log s - 1)$ defined in the interval $s > 0$ attains its minimum -1 at $s = 1$, we obtain from (4.3) the following estimates:

$$-|\Omega| \le L(t) \le L(0),$$

$$\int_0^t \int_\Omega q^{-1}|\nabla q|^2\, dxdt \le C_2 = L(0) + |\Omega|,$$

$$-\int_\Omega \varphi'(v)|\nabla v|^2\, dx \le 2C_2, \qquad (4.4)$$

where $|\Omega|$ stands for the volume of Ω. The second inequality of (4.4) reads

$$\|\nabla q^{1/2}\|_{L^2(0,t;L^2(\Omega))} \le C_2^{1/2}/2,$$

while

$$\|q(t)\|_1 = \|q_0\|_1 \qquad (4.5)$$

follows from $q \geq 0$ and the fact that

$$\frac{d}{dt} \int_\Omega q \, dx = 0.$$

Thus we obtain

$$\sup_{s \in (0,t)} \|p(s)\|_2 + \left\{ \int_0^t \|\nabla p(s)\|_2^2 \, ds \right\}^{1/2} \leq C_3, \quad p = q^{1/2}. \tag{4.6}$$

Henceforth $\| \cdot \|_r$ denotes the standard norm in $L^r(\Omega)$ unless otherswise stated.
   By (4.2), (4.6), and

$$0 \leq \frac{1}{|\Omega|} \int_\Omega q^{1/2} \, dx \leq \left(\frac{1}{|\Omega|} \int_\Omega q \, dx \right)^{1/2} = \left(\frac{1}{|\Omega|} \int_\Omega q_0 \, dx \right)^{1/2},$$

we obtain

$$\|q\|_{L^2(0,t;L^\infty(\Omega))} \leq C_4. \tag{4.7}$$

Now we show the following lemma.

Lemma 4.1.3 *It holds that*

$$\|q\|_{L^r(0,t;L^\infty(\Omega))} \leq C_5(r), \quad \|q\|_{L^r(\Omega \times (0,t))} \leq C_5(r) t^{1/r} \tag{4.8}$$

for each $2 \leq r < \infty$.

Proof Since $q(x, t) \geq 0$ implies

$$v(x, t) \geq v_0(x) \qquad \text{in } Q_T = \Omega \times (0, T), \tag{4.9}$$

the function $\varphi'(v(x, t))$ is uniformly bounded on $\overline{Q_T}$ because $\varphi' \leq 0 \leq (\varphi')'$, so that

$$\left| \varphi'(v(x, t)) \right| \leq C_6. \tag{4.10}$$

Given a smooth function $A = A(q)$, we have

$$\frac{d}{dt} \int_\Omega A(q) \, dx = \int_\Omega A'(q) q_t \, dx = - \int_\Omega A''(q) \nabla q \cdot (\nabla q - q \nabla \varphi(v)) \, dx,$$

which means

$$\frac{d}{dt} \int_\Omega A(q) \, dx + \int_\Omega a(q) |\nabla q|^2 \, dx = \int_\Omega a(q) q \nabla q \cdot \nabla \varphi(v) \, dx$$

with $a = A''$. Here it holds that

$$\int_\Omega a(q)q\nabla q \cdot \nabla \varphi(v)\, dx = \int_\Omega a(q)q(-\varphi'(v))^{1/2}(-\varphi'(v))^{1/2}\nabla q \cdot \nabla v\, dx$$

$$\leq \left\{ \int_\Omega -\varphi'(v)|\nabla v|^2\, dx \right\}^{1/2} \cdot \left\{ \int_\Omega a(q)^2 q^2(-\varphi'(v))|\nabla q|^2\, dx \right\}^{1/2}$$

$$\leq (2C_2C_6)^{1/2}\|a(q)^{1/2}q\|_\infty \cdot \left\{ \int_\Omega a(q)|\nabla q|^2\, dx \right\}^{1/2}$$

$$\leq 2C_2C_6\|a(q)^{1/2}q\|_\infty^2 + \frac{1}{4}\int_\Omega a(q)|\nabla q|^2\, dx,$$

and therefore,

$$\frac{d}{dt}\int_\Omega A(q)\, dx + \frac{3}{4}\int_\Omega a(q)|\nabla q|^2\, dx \leq 2C_2C_6\|a(q)^{1/2}q\|_\infty^2. \tag{4.11}$$

We thus obtain

$$\int_\Omega A(q)\, dx + \frac{3}{4}\int_0^t \int_\Omega a(q)|\nabla q|^2\, dxds$$

$$\leq \int_\Omega A(q_0)\, dx + 2C_2C_6 \int_0^t \|a(q)^{1/2}q\|_\infty^2\, ds. \tag{4.12}$$

First, if we take $A(q) = q^2$ in (4.12), we get

$$\|q(t)\|_2^2 + \frac{3}{2}\|\nabla q\|_{L^2(0,t;L^2(\Omega))}^2 \leq \|q_0\|_2^2 + 4C_2C_6\|q\|_{L^2(0,t;L^\infty(\Omega))}^2.$$

Then (4.7) implies

$$\sup_{s\in(0,t)} \|p(s)\|_2 + \left\{ \int_0^t \|\nabla p(s)\|_2^2\, ds \right\}^{1/2} \leq C_7, \quad p = q, \tag{4.13}$$

and hence

$$\|p\|_{L^2(0,t;L^\infty(\Omega))} \leq C_8, \quad p = q^2, \tag{4.14}$$

by (4.2). Next, putting $A(q) = q^{2m}$ in (4.12), we obtain

$$\|q^m(t)\|_2^2 + \frac{3}{2m} \cdot (2m-1) \cdot \|\nabla q^m(t)\|_{L^2(0,t;L^2(\Omega))}^2$$

$$\leq \|q_0^m\|_2^2 + 4m \cdot (2m-1) \cdot 2C_1C_6 \cdot \|q^m\|_{L^2(0,t;L^\infty(\Omega))}, \tag{4.15}$$

where $m = 2, 3, \ldots$. This yields

$$\sup_{s \in (0,t)} \|p(s)\|_2 + \left\{ \int_0^t \|\nabla p(s)\|_2^2 ds \right\}^{1/2} \le C_9(m), \quad p = q^m \tag{4.16}$$

for $m = 2$ by (4.14), and then

$$\|p\|_{L^2(0,t;L^\infty(\Omega))} \le C_{10}(m), \quad p = q^m \tag{4.17}$$

follows from (4.2) for $m = 2$. Continuing this process, we obtain (4.16) and (4.17) for any $m = 2^k$ ($k = 1, 2, \ldots$). This implies (4.8) for $r = 2^k$ ($k = 1, 2, \ldots$). Hence these inequalities hold for each $2 \le r < \infty$.   □

Putting

$$p = \int_0^t q \, ds = v - v_0, \tag{4.18}$$

we have

$$\Delta p = \int_0^t \Delta q \, ds = \int_0^t \left[q_t + \nabla \cdot (q \nabla \varphi(v)) \right] ds$$

$$= q - q_0 + \int_0^t \left[\varphi'(v) \nabla q \cdot \nabla v + q \varphi'(v) \Delta v + q \varphi''(v) |\nabla v|^2 \right] ds$$

$$= q - q_0 + \int_0^t \left[\varphi'(v) \nabla q \cdot (\nabla p + \nabla v_0) + q \varphi'(v) (\Delta p + \Delta v_0) \right.$$

$$\left. + q \varphi''(v) |\nabla p + \nabla v_0|^2 \right] ds = \psi - h,$$

with

$$\psi = \int_0^t q \varphi'(v) \Delta p \, ds \tag{4.19}$$

and

$$h = q_0 - q - \nabla v_0 \cdot \int_0^t \varphi'(v) \nabla q \, ds - \Delta v_0 \int_0^t q \varphi'(v) \, ds$$

$$- \int_0^t \left[\varphi'(v) \nabla q \cdot \nabla p + q \varphi''(v) |\nabla p + \nabla v_0|^2 \right] ds. \tag{4.20}$$

We conclude that

$$\psi_t = a \Delta p = a \psi - a h, \quad a = q \varphi'(v),$$

and hence

$$\psi(x, t) = -\int_0^t e^{A(x,t)-A(x,s)} a(x, s) h(x, s) \, ds$$

$$A(x, t) = \int_0^t a(x, t') dt'.$$

Thus we obtain

$$-\Delta p(x, t) = h(x, t) + \int_0^t e^{A(x,t)-A(x,s)} a(x, s) h(x, s) \, ds \qquad (4.21)$$

with

$$0 \le e^{A(x,t)-A(x,s)} \le 1, \qquad (4.22)$$

because $A_t = a = q\varphi'(v) \le 0$.

Lemma 4.1.4 *It holds that*

$$\|h\|_{L^\infty(0,t;L^1(\Omega))} \le C_{11}\left(1 + t^{3/2}(1 + K(t))\right), \qquad (4.23)$$

where

$$K(t) = \sup_{|z| \le \|v_0\|_\infty + C_4 t^{1/2}} |\varphi''(z)|.$$

Proof First we have

$$|v(x, t)| \le |v_0(x)| + \int_0^t q(x, \cdot) \le \|v_0\|_\infty + t^{1/2}\|q\|_{L^2(0,t;L^\infty(\Omega))}, \qquad (4.24)$$

and hence

$$\left|\varphi''(v(x, t))\right| \le K(t), \qquad (4.25)$$

by (4.7). Next, it holds that

$$\|q\|_{L^1(0,t;L^1(\Omega))} \le |\Omega| \, \|q\|_{L^1(0,t;L^\infty(\Omega))} \le C_{12} t^{1/2}, \qquad (4.26)$$

by (4.8) with $r = 2$. These inequalities, combined with (4.10), imply

$$\|h(t)\|_1 \le C_{13}\big(1 + t^{1/2} + t^{1/2}K(t) + \|\nabla q\|_{L^1(0,t;L^1(\Omega))}$$
$$+ \|\nabla q \cdot \nabla p\|_{L^1(0,t;L^1(\Omega))} + K(t)\|q|\nabla p|^2\|_{L^1(0,t;L^1(\Omega))}\big), \qquad (4.27)$$

because

$$|\nabla p + \nabla v_0|^2 \le 2(|\nabla p|^2 + |\nabla v_0|^2).$$

Here we apply (4.13) to obtain

$$\|\nabla q\|_{L^1(0,t;L^1(\Omega))} \le |\Omega|^{1/2} \|\nabla q\|_{L^1(0,t;L^2(\Omega))} \le C_{14}t^{1/2}. \qquad (4.28)$$

Furthermore,

$$
\begin{aligned}
\|q|\nabla p|^2\|_{L^1(0,t;L^1(\Omega))} &= \int_0^t \int_\Omega q(x,s) \left| \int_0^s \nabla q(x,\tau)d\tau \right|^2 dxds \\
&\le \int_0^t \int_\Omega sq(x,s) \left\{ \int_0^s |\nabla q(x,\tau)|^2 d\tau \right\} dxds \\
&\le t\|\nabla q\|^2_{L^2(0,t;L^2(\Omega))} \int_0^t \|q(s)\|_\infty ds \\
&\le t^{3/2}\|q\|_{L^2(0,t;L^\infty(\Omega))}\|\nabla q\|^2_{L^2(0,t;L^2(\Omega))} \\
&\le C_4 C_7^2 t^{3/2} \qquad\qquad\qquad\qquad\qquad (4.29)
\end{aligned}
$$

and

$$
\begin{aligned}
\|\nabla q \cdot \nabla p\|_{L^1(0,t;L^1(\Omega))} &= \|q^{-1/2}\nabla q \cdot q^{1/2}\nabla p\|_{L^1(0,t;L^1(\Omega))} \\
&\le \|q^{-1}|\nabla q|^2\|^{1/2}_{L^1(0,t;L^1(\Omega))} \|q|\nabla p|^2\|^{1/2}_{L^1(0,t;L^1(\Omega))} \\
&\le C_1^{1/2} \cdot C_4^{1/2}C_7 \cdot t^{3/4}. \qquad\qquad\qquad (4.30)
\end{aligned}
$$

Inequalities (4.27)–(4.30) now imply (4.23).   □

Lemma 4.1.5 *It holds that*

$$
\begin{aligned}
\|\varphi(v)\|_{L^\infty(0,t;L^\infty(\Omega))} &\le k(t), \\
\|\Delta\varphi(v)\|_{L^\infty(0,t;L^1(\Omega))} &\le C_{15}\big(1 + t^2 + (1 + t^{5/2})K(t)^2\big) \qquad (4.31)
\end{aligned}
$$

where

$$k(t) = \sup_{|z| \le \|v_0\|_\infty + C_4 t^{1/2}} |\varphi(z)|.$$

Proof The first inequality in (4.31) follows from (4.7) and (4.24).

To show the second inequality, we use p, ψ, and h as defined by (4.18), (4.19), and (4.20), respectively. It holds that

$$|\Delta\varphi(v)| \le C_{16}\big(1 + |\Delta p| + K(t)(1 + |\nabla p|^2)\big), \qquad (4.32)$$

thanks to (4.10), (4.25), and

$$\Delta\varphi(v) = \varphi'(v)(\Delta p + \Delta v_0) + \varphi''(v)|\nabla p + \nabla v_0|^2.$$

Then we apply the Gagliardo-Nirenberg inequality valid for $N = 1$, to obtain

$$
\begin{aligned}
\|\nabla p(t)\|_2^2 &\leq C_{17} \|p(t)\|_{W^{2,1}(\Omega)} \|p(t)\|_{L^\infty(\Omega)} \\
&\leq C_{17}(\|\Delta p(t)\|_1 + \|p(t)\|_1) \cdot \|q\|_{L^1(0,t;L^\infty(\Omega))} \\
&\leq C_{17}(\|\Delta p(t)\|_1 + \|q\|_{L^1(0,t;L^1(\Omega))}) \cdot \|q\|_{L^1(0,t;L^\infty(\Omega))} \\
&\leq C_{17}C_4 t^{1/2}(|\Omega| C_4 t^{1/2} + \|\Delta p(t)\|_1).
\end{aligned}
$$

It follows that

$$
\|\Delta \varphi(v)(t)\|_1 \leq C_{18}\left\{ 1 + \|\Delta p(t)\|_1 + K(t)\left(1 + \|\nabla p(t)\|_2^2\right)\right\}. \tag{4.33}
$$

Since

$$
\|\Delta p(t)\|_1 \leq \|\psi(t)\|_1 + \|h(t)\|_1
$$

and

$$
\|h\|_{L^\infty(0,t;L^1(\Omega))} \leq C_{11}\left(1 + t^{3/2}(1 + K(t))\right),
$$

inequality (4.22) implies that

$$
\begin{aligned}
\|\psi(t)\|_1 &\leq \int_0^t \|a(s)h(s)\|_1 ds \leq C_6 \int_0^t \|q(s)h(s)\|_1 ds \\
&\leq C_6 \int_0^t \|q(s)\|_\infty \|h(s)\|_1 ds \\
&\leq C_6 \|h\|_{L^\infty(0,t;L^1(\Omega))} \|q\|_{L^1(0,t;L^\infty(\Omega))} \\
&\leq C_6 C_{11}\left(1 + t^{1/2}(t + K(t))\right)t^{1/2}\|q\|_{L^2(0,t;L^\infty(\Omega))}. \tag{4.34}
\end{aligned}
$$

Then we obtain

$$
\begin{aligned}
\|\Delta p(t)\|_1 &\leq C_{19}t^{1/2}\left\{ 1 + t^2 + (1 + t)K(t)\right\}, \\
\|\nabla p(t)\|_2^2 &\leq C_{19}t^{1/2}\left(1 + t^2(1 + K(t))\right). \tag{4.35}
\end{aligned}
$$

Inequalities (4.33) and (4.35) imply (4.31), because $K(t) \geq K(0) > 0$. $\qquad\square$

In the following, $m_{i,r,T}(t)$, $i = 1, \ldots, 4$, denote bounded functions of $t \in (0, T]$ satisfying

$$
\lim_{t \downarrow 0} m_{i,r,T}(t) = 0,
$$

where $1 \leq r < \infty$.

Lemma 4.1.6 *We have*

$$
\|h\|_{L^r(\Omega \times (0,t))} \leq m_{1,r,T}(t)\left(1 + \|\nabla q\|_{L^1(0,t;L^r(\Omega))}\right) \tag{4.36}
$$

for each $1 \leq r < \infty$.

Proof By (4.20), it holds that

$$\|h\|^r_{L^r(\Omega\times(0,t))} \le C_{20}(r)\Big\{ t + \|q\|^r_{L^r(\Omega\times(0,t))} + \int_0^t \Big(\int_0^s \|\nabla q(\tau)\|_r d\tau \Big)^r ds$$

$$+ \int_0^t \Big(\int_0^s \|\nabla q(\tau)\cdot\nabla p(\tau)\|_r d\tau \Big)^r ds + (1+K(t)) \int_0^t \Big(\int_0^s \|q(\tau)\|_r d\tau \Big)^r ds$$

$$+ K(t) \int_0^t \Big(\int_0^s \|q(\tau)|\nabla p(\tau)|^2\|_r d\tau \Big)^r ds \Big\}. \tag{4.37}$$

Here we have

$$\|q\|^r_{L^r(\Omega\times(0,t))} \le C_{21}(r)t \tag{4.38}$$

and

$$\int_0^t \Big(\int_0^s \|q(\tau)\|_r d\tau \Big)^r ds \le \int_0^t s^{r-1} \|q(s)\|^r_r ds$$

$$\le t^{r-1} \|q\|^r_{L^r(Q_T)} \le C_{21}(r)t^r, \tag{4.39}$$

by Lemma 4.1.3, recalling that $Q_T = \Omega \times (0,T)$. We also have

$$\int_0^t \Big(\int_0^s \|q(\tau)|\nabla p(\tau)|^2\|_r d\tau \Big)^r ds \le \int_0^t \Big(\int_0^s \|q(\tau)\|_r \|\nabla p(\tau)\|^2_\infty d\tau \Big)^r ds$$

$$\le \|\nabla p\|^{2r}_{L^\infty(\Omega\times(0,t))} \int_0^t s^{r-1} \|q\|^r_{L^r(\Omega\times(0,s))} ds$$

$$\le t^r \|\nabla p\|^{2r}_{L^\infty(\Omega\times(0,t))} \|q\|^r_{L^r(\Omega\times(0,t))} \le C_{22}(r)t^{r+1} \|\nabla p\|^{2r}_{L^\infty(\Omega\times(0,t))}.$$

Then (4.26), (4.35), and the embedding

$$W^{2,1}(\Omega) \subset C^1(\overline{\Omega}), \tag{4.40}$$

valid for $N = 1$, imply that

$$\|\nabla p(t)\|_\infty \le C_{23}(1+t^2)(1+K(t)).$$

We thus end up with the inequality

$$\int_0^t \Big(\int_0^s \|q(\tau)|\nabla p(\tau)|^2\|_r d\tau \Big)^r ds \le C_{24}(r)t^{r+1}(1+t^{4r})(1+K(t)^{2r}). \tag{4.41}$$

Moreover,

$$\int_0^t \Big(\int_0^s \|\nabla q(\tau) \cdot \nabla p(\tau)\|_r d\tau \Big)^r ds$$

$$\leq \|\nabla p\|_{L^\infty(\Omega \times (0,t))}^r \int_0^t \Big(\int_0^s \|\nabla q(\tau)\|_r d\tau \Big)^r ds$$

$$\leq \|\nabla p\|_{L^\infty(\Omega \times (0,t))}^r \cdot t \|\nabla q\|_{L^1(0,t;L^r(\Omega))}^r. \tag{4.42}$$

Inequality (4.36) now follows from (4.37)–(4.42). $\qquad\square$

Lemma 4.1.7 *It holds that*

$$\|\Delta \varphi(v)\|_{L^r(\Omega \times (0,t))} \leq m_{2,r,T}(t)\big(1 + \|h\|_{L^r(\Omega \times (0,t))}\big) \tag{4.43}$$

for $1 < r < \infty$.

Proof First, we have

$$\|\Delta \varphi(v)\|_{L^r(\Omega \times (0,t))} \leq C_{25}(r)\Big(t^{1/r}\big(1 + K(t)\big)$$

$$+ K(t)\|\nabla p\|_{L^{2r}(\Omega \times (0,t))}^2 + \|\Delta p\|_{L^r(\Omega \times (0,t))}\Big), \tag{4.44}$$

thanks to (4.32). Next, the Gagliardo-Nirenberg inequality and Lemma 4.1.3 imply

$$\|\nabla p(s)\|_{2r}^{2r} \leq C_{26}(r)\|p(s)\|_{W^{2,r}(\Omega)}^r \|p(s)\|_\infty^r$$

$$\leq C_{26}(r)\|q\|_{L^1(0,s;L^\infty(\Omega))}^r \big(\|\Delta p(s)\|_r^r + \|p(s)\|_r^r\big)$$

$$\leq C_{26}(r)C_4^r s^{r/2}\big(\|\Delta p(s)\|_r^r + \|q\|_{L^1(0,s;L^\infty(\Omega))}^r\big)$$

$$\leq C_{27}(r)s^{r/2}\big(s^{r/2} + \|\Delta p(s)\|_r^r\big),$$

which implies

$$\|\nabla p\|_{L^{2r}(\Omega \times (0,t))}^2 \leq C_{27}(r)\big\{t^{1+1/r} + t^{1/2}\|\Delta p\|_{L^r(\Omega \times (0,t))}\big\}. \tag{4.45}$$

Finally, we have

$$\|\psi(t)\|_r \leq C_6 \|q\|_{L^{r'}(0,t;L^\infty(\Omega))} \cdot \Big(\int_0^t \|h(s)\|_r^r ds \Big)^{1/r}$$

for $\frac{1}{r'} + \frac{1}{r} = 1$, and hence

$$\|\Delta p\|_{L^r(\Omega \times (0,t))} \leq \|h\|_{L^r(\Omega \times (0,t))} + \|\psi\|_{L^r(\Omega \times (0,t))}$$

$$\leq \big(1 + C_6 t^{1/r}\|q\|_{L^{r'}(0,t;L^\infty(\Omega))}\big)\|h\|_{L^r(\Omega \times (0,t))} \tag{4.46}$$

Inequality (4.43) now follows from (4.44)–(4.46). $\qquad\square$

We are ready to prove Theorem 4.1, using the norms below (see [76]):

$$\|q\|_{W^{2,1}_r(Q_T)} \equiv \sum_{0 \le 2\alpha+\beta \le 2} \|D^\alpha_t D^\beta_x q\|_{L^r(Q_T)},$$

$$[q]_{1+\alpha, Q_T} \equiv \|q_x\|_\infty + \|q\|_\infty + \sup_{x \ne y, \, t} \frac{|q_x(x,t) - q_x(y,t)|}{|x-y|^\alpha}$$

$$+ \sup_{t \ne s, \, x} \frac{|q(x,t) - q(x,s)|}{|t-s|^{\alpha/2}},$$

$$[f]_{\alpha, Q_T} = \|f\|_\infty + \sup_{x \ne y, \, t} \frac{|f(x,t) - f(y,t)|}{|x-y|^\alpha} + \sup_{t \ne s, \, x} \frac{|f(x,t) - f(x,s)|}{|t-s|^{\alpha/2}},$$

$$[f]_{2+\alpha, Q_T} = \sum_{0 \le 2\alpha+\beta \le 2} \|D^\alpha_t D^\beta_x f\|_\infty + \sup_{x \ne y, \, t} \frac{|f_{xx}(x,t) - f_{xx}(y,t)|}{|x-y|^\alpha}$$

$$+ \sup_{t \ne s, \, x} \frac{|f_t(x,t) - f_t(x,s)|}{|t-s|^{\alpha/2}}.$$

Proof of Theorem 4.1 First, (4.13) implies

$$\|\nabla q\|_{L^2(\Omega \times (0,t))} \le C_7,$$

and hence

$$\|\nabla q\|_{L^1(0,t;L^2(\Omega))} \le C_7 t^{1/2}.$$

Next, we can apply Lemmas 4.1.6 and 4.1.7 for $r = 2$ to get

$$\|\Delta\varphi(v)\|_{L^2(\Omega \times (0,t))} \le m_{3,2,T}(t). \tag{4.47}$$

On the other hand,

$$\|\nabla\varphi(v)(t)\|_\infty \le C_{28}\left\{1 + t^{5/2} + k(t) + (1+t^3)K(t)^2\right\}$$

by Lemma 4.1.5 and (4.40), and then it follows that

$$\|\nabla\varphi(v)\|_{L^\infty(\Omega \times (0,t))} \le C_{29}(t). \tag{4.48}$$

Here we apply the parabolic L^2-regularity (see Theorem II.3 of [106]) to

$$q_t = \Delta q - \nabla q \cdot \nabla\varphi(v) - q\Delta\varphi(v) \qquad \text{in } \Omega \times (0,T),$$

$$\frac{\partial q}{\partial \nu} = 0 \qquad\qquad\qquad\qquad \text{on } \Omega \times (0,T). \tag{4.49}$$

From the estimates (4.47) and (4.48) it follows that

$$\|q\|_{L^2(0,T;H^2(\Omega))} \leq C_{30}(T),$$

which implies

$$\|\nabla q\|_{L^1(0,T;L^r(\Omega))} \leq C_{31}(r,T)$$

for $r > 1$ by $N = 1$. Then by Lemmas 4.1.6 and 4.1.7 we obtain

$$\|\Delta\varphi(v)\|_{L^r(\Omega\times(0,t))} \leq m_{4,r,T}(t). \tag{4.50}$$

Now we apply (4.50) and (4.48) to (4.49). We obtain

$$\|q\|_{W_r^{2,1}(Q_T)} \leq C_{32}(r,T),$$

thanks to the parabolic L^r-regularity.

Then the embedding

$$W_r^{2,1}(Q_T) \hookrightarrow C^{2-(N+2)/r,\ 1-(N+2)/2r}(\overline{Q_T}),$$

valid to $r > N + 2$, $N = 1$, is available. It holds that

$$[q]_{1+\alpha,Q_T} \leq C_{33}(T) \tag{4.51}$$

for $0 < \alpha < 1$, whence

$$[v]_{1+\alpha,Q_T} \leq C_{34}(T)$$

since $v_t = q$. Furthermore,

$$[p_{xx}]_{\alpha,Q_T} + [p_x]_{\alpha,Q_T} \leq C_{35}(T) \tag{4.52}$$

by (4.21). Then (4.18) yields

$$[\varphi(v)]_{2+\alpha,Q_T} \leq C_{36}(T).$$

The parabolic Schauder estimate now implies

$$[q]_{2+\alpha,Q_T} \leq C_{37}(T). \tag{4.53}$$

By the construction of the local-in-time solution (see [129]), inequality (4.53) guarantees its extension global-in-time. $\qquad\square$

4.2 Parabolic Systems with Non-local Term

The mathematical analysis of (3.34) was carried out in [62, 63]. It has determined the decay rate as $t \uparrow +\infty$ and displayed the phase separation $k \uparrow +\infty$ in accordance with the non-local case $R = 0$ studied by [49, 50, 52], that is,

$$\frac{\partial q_A}{\partial t} = D_A \Delta q_A - k_A q_A q_B,$$

$$\frac{\partial q_B}{\partial t} = D_B \Delta q_B - k_B q_A q_B. \qquad (4.54)$$

Here we take the system of parabolic equations with non-local term

$$\frac{\partial q_A}{\partial t} = D_A \Delta q_A - \frac{k_A}{\omega_N R^N} \int_{B(\cdot,R) \cap \Omega} q_B dy \cdot q_A,$$

$$\frac{\partial q_B}{\partial t} = D_B \Delta q_B - \frac{k_B}{\omega_N R^N} \int_{B(\cdot,R) \cap \Omega} q_A dy \cdot q_B \quad \text{in } \Omega \times (0, T),$$

subject to the initial and boundary conditions

$$\frac{\partial q_A}{\partial \nu} = \frac{\partial q_B}{\partial \nu} = 0 \qquad\qquad \text{on } \partial\Omega \times (0, T),$$

$$q_A|_{t=0} = q_{0A}(x) \geq 0, \quad q_B|_{t=0} = q_{0B}(x) \geq 0 \quad \text{in } \Omega,$$

where $\Omega \subset \mathbf{R}^N$ and ν are a bounded domain with smooth boundary $\partial\Omega$ and its outer normal unit vector, respectively. Since

$$0 \leq q_A = [A]/[A]_* \leq 1, \quad k_A = k[B]_*,$$

$$0 \leq q_B = [B]/[B]_* \leq 1, \quad k_B = k[A]_*,$$

we have

$$u_t = d_1 \Delta u - \frac{ku}{\omega_N R^N} \int_{B(\cdot,R) \cap \Omega} v \, dy,$$

$$v_t = d_2 \Delta v - \frac{kv}{\omega_N R^N} \int_{B(\cdot,R) \cap \Omega} u \, dy \quad \text{in } \Omega \times (0, T), \qquad (4.55)$$

and

$$\frac{\partial u}{\partial \nu} = \frac{\partial v}{\partial \nu} = 0 \qquad\qquad \text{on } \partial\Omega \times (0, T),$$

$$u|_{t=0} = u_0(x) \geq 0, \quad v|_{t=0} = v_0(x) \geq 0 \quad \text{in } \Omega, \qquad (4.56)$$

using the relations

$$u = [A] = q_A[A]_*, \quad d_1 = D_A,$$
$$v = [B] = q_B[B]_*, \quad d_2 = D_B.$$

By a standard argument we obtain the existence and uniqueness of a local-in-time classical solution

$$(u, v) \in C^{2,1}(\overline{\Omega} \times [0, T))^2$$

to (4.55)–(4.56) for $0 < T \ll 1$, under appropriate assumptions, say,

$$0 \le u_0, v_0 \in C^2(\overline{\Omega}), \quad \frac{\partial u_0}{\partial \nu} = \frac{\partial v_0}{\partial \nu} = 0 \quad \text{on } \partial\Omega. \tag{4.57}$$

Then the comparison theorem yields

$$0 \le u \le \|u_0\|_\infty, \ 0 \le v \le \|v_0\|_\infty \quad \text{in } Q_T = \Omega \times (0, T). \tag{4.58}$$

By this a priori estimate and the standard parabolic regularity, the solution extends globally in time [76].

Using L^p and Schauder estimates, we obtain

$$\|u\|_{C^{1+\theta,1/2+\theta/2}(Q_T)} + \|v\|_{C^{1+\theta,1/2+\theta/2}(Q_T)} \le C$$

for $0 < \theta < 1$, with a constant $C = C(k, T)$ independent of $R > 0$, which guarantees the convergence as $R \downarrow 0$,

$$u_R \longrightarrow u, \quad v_R \longrightarrow v \quad \text{in } C^{1,0}(\overline{Q_T}), \tag{4.59}$$

passing to a subsequence. This $(u, v) \in C^{1,0}(\overline{Q_T})$ is a weak solution to

$$u_t = d_1 \Delta u - kuv,$$
$$v_t = d_2 \Delta v - kvu \quad \text{in } \Omega \times (0, T), \tag{4.60}$$

satisfying (4.56), and hence is the classical solution, again by the parabolic regularity. Then we obtain (4.59) as $R \downarrow 0$ from the uniqueness of this solution without taking subsequences.

Several properties are known for system (4.60) with conditions (4.56). First, one can compute the decay rate [49, 50, 52]. If $\|u_0\|_1 = \|v_0\|_1$, this rate is subject to the ODE part

$$\frac{dU}{dt} = -kUV, \quad U(0) = \bar{u}_0 \equiv \frac{1}{|\Omega|} \int_\Omega u_0 \, dx,$$

$$\frac{dV}{dt} = -kUV, \quad V(0) = \bar{v}_0 \equiv \frac{1}{|\Omega|} \int_\Omega v_0 \, dx.$$

Then it follows that

$$U(t) = V(t) = \frac{1}{kt + U(0)^{-1}}.$$

Letting $t \uparrow +\infty$, we obtain

$$\|(u, v)(\cdot, t)\|_\infty = O(t^{-1}),$$
$$\|(u, v)(\cdot, t) - (U, V)(t)\|_\infty = O(t^{-2}),$$
$$\|(u, v)(\cdot, t) - (\bar{u}, \bar{v})(t)\|_\infty = O(e^{-d_0 \mu_2 t}), \tag{4.61}$$

with

$$\bar{u}(t) = \frac{1}{|\Omega|} \int_\Omega u(x, t) \, dx, \quad \bar{v}(t) = \frac{1}{|\Omega|} \int_\Omega v(x, t) \, dx$$

where

$$d_0 = \min\{d_1, d_2\} \tag{4.62}$$

and $\mu_2 > 0$ denotes the second eigenvalue of $-\Delta$ with the Neumann boundary condition. In the other case, say $\|u_0\|_1 > \|v_0\|_1$, only u survives as $t \uparrow +\infty$,

$$\|u(\cdot, t) - u_\infty\|_\infty = \begin{cases} O(e^{-\beta t}), & d_1 \mu_2 \neq u_\infty, \\ O(te^{-\beta t}), & d_1 \mu_2 = u_\infty, \end{cases}$$
$$\|v(\cdot, t)\|_\infty = O(e^{-u_\infty t}), \tag{4.63}$$

where $\beta = \min\{d_1 \mu_2, u_\infty\}$ and

$$u_\infty = \frac{1}{|\Omega|} \int_\Omega (u_0 - v_0) \, dx > 0. \tag{4.64}$$

It is reasonable in the context of chemistry that if the initial concentrations $[A]$ and $[B]$ of the A and B molecules are equal, then they eventually vanish. Otherwise, only the reactant with the biggest concentration will survive. The difference of decay rates described above, however, is to be noted, for example, in the pathway modeling of Sect. 1.1.

The second property of (4.60) associated with (4.56) is the phase separation as $k \uparrow +\infty$ (see [32]). Let $z = -u$ and assume $z_0 \cdot v_0 = 0$, $z_0 = -u_0$. Then (z^k, v^k) converges strongly in $L^1(Q_T)$; denote the limit by (z, v). Here $w = z + v$ is the solution to

$$w_t = \nabla \cdot d(w)\nabla w \qquad \text{in } \Omega \times (0, T),$$

$$\frac{\partial w}{\partial \nu} = 0 \qquad \text{on } \partial\Omega \times (0, T),$$

$$w|_{t=0} = w_0(x) \qquad \text{in } \Omega, \tag{4.65}$$

where $w_0 = z_0 + v_0$ and

$$d(w) = \begin{cases} d_1, & \text{if } w < 0 \\ \frac{d_1+d_2}{2}, & \text{if } w = 0 \\ d_2, & \text{if } w > 0. \end{cases}$$

Furthermore, it holds that

$$w^+ = v, \quad w^- = -z, \quad w^+ \cdot w^- = 0$$

for $w^{\pm} = \max\{\pm w, 0\}$. Interaction of the A and B molecules thus separates into two phases as $k \uparrow +\infty$, provided that $z_0 \cdot v_0 = 0$.

More precisely, this $w = w(x, t) \in L^1(Q_T)$ is a weak solution satisfying $\nabla w \in L^2(Q_T)$ and

$$\iint_{Q_T} [w\xi_t - d(w)\nabla w \cdot \nabla \xi] \, dxdt + \int_{\Omega} w_0(x)\xi(x, 0)dx = 0 \tag{4.66}$$

for any

$$\xi = \xi(x, t) \in C^1(\overline{\Omega} \times [0, T)), \quad \xi = 0, \ 0 < T - t \ll 1. \tag{4.67}$$

The value $d(w) = \frac{d_1+d_2}{2}$ at $w = 0$ does not contribute in (4.66) because $\nabla w = 0$ a.e. on $\{w = 0\}$. It is also the solution to the free boundary problem studied in [17], and hence $w = w(x, t) \in C^{\theta, \theta/2}(\overline{Q_T}), 0 < \theta < 1$. Partial regularity also holds [136]. There is thus an open set $\mathcal{O}$ in Q_T such that $w = w(x, t) \in C^{2+\theta, 1+\theta/2}(\mathcal{O})$ and the part $\{(x, t) \in \mathcal{O} \mid w(x, t) = 0\}$ of the free boundary is $C^{2+\theta, 1+\theta/2}$. The residual set $W = Q_T \setminus \mathcal{O}$, furthermore, is composed of W_i, $i = 1, 2$, satisfying $\mathcal{P}^N(W_1) = 0$ and

$$\lim_{r \downarrow 0} \frac{1}{r^{N+2}} \int_{P_r(x,t)} |\nabla w|^2 dxdt = 0, \quad (x, t) \in W_2,$$

where

$$P_r(x, t) = \{(y, s) \in Q_T \mid |y - x| < r, \ |s - t| < r^2\}$$

$$\mathcal{P}^N(W) = \liminf_{\delta \downarrow 0} \left\{ \sum_j r_j^N \mid W \subset \bigcup_j P_{r_j}(x_j, t_j), \ 2r_j < \delta \text{ for all } j \right\}.$$

Finally, there is a phase field model associated to (4.65):

$$w_t = \nabla \cdot d(\varphi)\nabla w$$
$$\tau\varphi_t = -\delta\mathcal{F}_w(\varphi) \qquad\qquad \text{in } \Omega \times (0, T),$$
$$\frac{\partial w}{\partial\nu} = \frac{\partial\varphi}{\partial\nu} = 0 \qquad\qquad \text{on } \partial\Omega \times (0, T),$$
$$w|_{t=0} = w_0(x), \quad \varphi|_{t=0} = \varphi_0(x) \quad \text{in } \Omega,$$

where

$$d(\varphi) = \frac{1}{2}(d_1 - d_2)\varphi + \frac{1}{2}(d_1 + d_2),$$

$$\mathcal{F}_w(\varphi) = \int_\Omega \left[\frac{\xi^2}{2}|\nabla\varphi|^2 + W(\varphi) - 2w\varphi \right] dx$$

$$W(\varphi) = \frac{1}{4}(1 - \varphi^2)^2.$$

Here, ξ is a constant related to the molecular distance. This phase field model actually realizes the continuously varying diffusion coefficient which takes the value d_1, $\frac{d_1+d_2}{2}$, and d_2 according to whether $\varphi = 1$, $\varphi = 0$, and $\varphi = -1$, respectively. Conversely, $\varphi = 1$ and $\varphi = -1$ are stable according to whether $w > 0$ and $w < 0$, respectively.

Here we study phase separation and asymptotic convergence. First, putting $z = -u$ in (4.55) and (4.56), we obtain

$$z_t = d_1\Delta z - \frac{z}{\varepsilon \cdot \omega_N R^N}\int_{B(\cdot,R)\cap\Omega} v\, dy,$$

$$v_t = d_2\Delta v + \frac{v}{\varepsilon \cdot \omega_N R^N}\int_{B(\cdot,R)\cap\Omega} z\, dy \qquad \text{in } \Omega \times (0, T) \qquad (4.68)$$

with

$$\frac{\partial z}{\partial\nu} = \frac{\partial v}{\partial\nu} = 0, \qquad\qquad \text{on } \partial\Omega \times (0, T)$$

$$z|_{t=0} = z_0(x) = -u_0(x) \le 0, \quad v|_{t=0} = v_0(x) \ge 0 \qquad \text{in } \Omega \qquad (4.69)$$

for $\varepsilon = k^{-1}$. If the initial states of the A and B molecules are separated as

$$v_0(x) \cdot z_0(y) = 0, \qquad |x - y| < R, \ x, y \in \Omega, \qquad\qquad (4.70)$$

then there exists the limit (z, v) of $(z^\varepsilon, v^\varepsilon)$ as $\varepsilon \downarrow 0$, which satisfies

$$v(x, t) \cdot z(y, t) = 0, \qquad |x - y| < R, \ x, y \in \Omega, \ t \ge 0.$$

Theorem 4.2 *If (z_0, v_0) satisfies (4.57) and (4.70), then any sequence $\varepsilon_j \downarrow 0$ admits a subsequence, denoted by the same symbol, such that*

$$v^{\varepsilon_j} \longrightarrow w^+, \quad z^{\varepsilon_j} \longrightarrow -w^-, \quad v^{\varepsilon_j} + z^{\varepsilon_j} \longrightarrow w \qquad in \ L^1(Q_T)$$
$$\nabla v^{\varepsilon_j} \longrightarrow \nabla w^+, \quad \nabla z^{\varepsilon_j} \longrightarrow -\nabla w^- \qquad in \ L^2(Q_T).$$

It holds that

$$w^+(x, t) \cdot w^-(y, t) = 0 \quad for \ a.e. \ x, y \in \Omega, \ |x - y| < R, t \geq 0.$$

A similar result arises for relaxed reaction radii and then the limit equation takes the form (4.65) (see [62]). Henceforth $C_i, i = 1, 2, \ldots, 11$, denote positive constants independent of ε and $(z, v) = (z^\varepsilon(x, t), v^\varepsilon(x, t))$ is the solution to (4.68)–(4.69).

Lemma 4.2.1 *It holds that*

$$\iint_{Q_T} \left[|\nabla z|^2 + |\nabla v|^2 \right] dxdt + \frac{1}{d_0\varepsilon} \iint_{Q_T} dxdt \times$$

$$\times \left[\frac{z^2(x, t)}{\omega_N R^N} \int_{B(x,R)\cap\Omega} v(y, t) \, dy - \frac{v^2(x, t)}{\omega_N R^N} \int_{B(x,R)\cap\Omega} z(y, t) \, dy \right] dxdt$$
$$\leq C_1 \tag{4.71}$$

with $d_0 > 0$ defined by (4.62).

Proof First, we have

$$\|z(\cdot, t)\|_\infty, \ \|v(\cdot, t)\|_\infty \leq C_2, \qquad 0 \leq t < T \tag{4.72}$$

by (4.58). Next we use the equalities

$$\int_\Omega \left[v^2(x, T) + z^2(x, T) - v_0^2(x) - z_0^2(x) \right] dx = \iint_{Q_T} \frac{\partial}{\partial t}(v^2 + z^2)dxdt$$

$$= 2 \iint_{Q_T} v \left[\left(d_2 \Delta v + \frac{v}{\varepsilon \cdot \omega_N R^N} \int_{B(\cdot,R)\cap\Omega} z(y, t) \, dy \right) \right.$$

$$+ z \left(d_1 \Delta z - \frac{z}{\varepsilon \cdot \omega_N R^N} \int_{B(\cdot,R)\cap\Omega} v(y, t) \, dy \right) \right] dxdt$$

to obtain

$$2d_2 \iint_{Q_T} |\nabla v|^2 dxdt + 2d_1 \iint_{Q_T} |\nabla z|^2 dxdt + \frac{2}{\varepsilon} \iint_{Q_T} dxdt \times$$

$$\times \left[\frac{z^2(x,t)}{\omega_N R^N} \int_{B(x,R)\cap\Omega} v(y,t)\, dy - \frac{v^2(x,t)}{\omega_N R^N} \int_{B(x,R)\cap\Omega} z(y,t)\, dy \right]$$

$$= \int_{\Omega} \left[v_0^2(x) + z_0^2(x) - v^2(x,T) - z^2(x,T) \right] dx \le 2|\Omega| C_2^2.$$

It holds that

$$\iint_{Q_T} \left(|\nabla v|^2 + |\nabla z|^2 \right) dxdt + \frac{1}{d_0\varepsilon} \iint_{Q_T} dxdt \times$$

$$\times \left[\frac{z^2(x,t)}{\omega_N R^N} \int_{B(x,R)\cap\Omega} v(y,t)\, dy - \frac{v^2(x,t)}{\omega_N R^N} \int_{B(x,R)\cap\Omega} z(y,t)\, dy \right]$$

$$\le \frac{|\Omega|}{d_0} C_2^2 = C_1.$$

$\square$

Lemma 4.2.2 *If (4.70) holds, then*

$$\|v_t(\cdot,t)\|_1 + \|z_t(\cdot,t)\|_1 \le C_3, \quad 0 \le t < T. \tag{4.73}$$

Proof Given a C^2 convex function $\Phi(x)$ such that $\Phi(0) = \Phi'(0) = 0$, we have

$$\int_0^T \frac{d}{dt} \int_{\Omega} [\Phi(v_t) + \Phi(z_t)]\, dxdt$$

$$= \iint_{Q_T} \Phi'(v_t) \left[d_2 \Delta v_t + \left(\frac{v}{\varepsilon \cdot \omega_N R^N} \int_{B(\cdot,R)\cap\Omega} z(y,t)\, dy \right)_t \right] dxdt$$

$$+ \iint_{Q_T} \Phi'(z_t) \left[d_1 \Delta z_t - \left(\frac{z}{\varepsilon \cdot \omega_N R^N} \int_{B(\cdot,R)\cap\Omega} v(y,t)\, dy \right)_t \right] dxdt$$

$$= \iint_{Q_T} \left[-d_2 \Phi''(v_t) |\nabla v_t|^2 - d_1 \Phi''(z_t) |\nabla z_t|^2 \right.$$

$$+ \Phi'(v_t) \left(\frac{v}{\omega_N R^N} \int_{B(\cdot,R)\cap\Omega} z(y,t)\, dy \right)_t$$

$$\left. - \Phi'(z_t) \left(\frac{z}{\omega_N R^N} \int_{B(\cdot,R)\cap\Omega} v(y,t)\, dy \right)_t \right] dxdt.$$

Since

$$\int_\Omega \Phi'(v_t)\Big(\frac{v}{\omega_N R^N}\int_{B(\cdot,R)\cap\Omega} z(y,t)\,dy\Big)_t dx$$

$$= \iint_{\mathbf{R}^N\times\mathbf{R}^N} dxdy\,\Phi'(v_t(x,t))\frac{\chi_{|x-y|<R}(x,y)}{\omega_N R^N}\chi_\Omega(x)\chi_\Omega(y)\,\{v(x,t)z(y,t)\}_t$$

and

$$\int_\Omega \Phi'(z_t)\Big(\frac{z}{|B(\cdot,R)|}\int_{B(\cdot,R)\cap\Omega} v(y,t)\,dy\Big)_t dx$$

$$= \iint_{\mathbf{R}^N\times\mathbf{R}^N} \Phi'(z_t(x,t))\frac{\chi_{|x-y|<R}(x,y)}{\omega_N R^N}\chi_\Omega(x)\chi_\Omega(y)\,\{v(y,t)z(x,t)\}_t\,dxdy$$

$$= \iint_{\mathbf{R}^N\times\mathbf{R}^N} \Phi'(z_t(y,t))\frac{\chi_{|x-y|<R}(x,y)}{\omega_N R^N}\chi_\Omega(x)\chi_\Omega(y)\,\{v(x,t)z(y,t)\}_t\,dxdy$$

it holds that

$$\int_\Omega [\Phi(v_t(x,T))+\Phi(z_t(x,T))]\,dx \le \int_\Omega [\Phi(v_t(x,0))+\Phi(z_t(x,0))]\,dx$$

$$+\frac{1}{\varepsilon}\int_0^T dt \iint_{\mathbf{R}^N\times\mathbf{R}^N} \frac{\chi_{|x-y|<R}(x,y)}{\omega_N R^N}\chi_\Omega(x)\chi_\Omega(y)\times$$

$$\times [\Phi'(v_t(x,t))-\Phi'(z_t(y,t))]\,[v_t(x,t)z(y,t)+v(x,t)z_t(y,t)]\,dxdy$$

$$(4.74)$$

because $\Phi'' \ge 0$. Now we replace Φ in (4.74) by a sequence of convex functions $\Phi_n = \Phi_n(s)$ such that $|\Phi_n'(s)| \le 1$,

$$\Phi_n(s) \to |s| \qquad \text{locally uniformly in } s \in \mathbf{R},$$
$$\Phi_n'(s) \to \operatorname{sgn}(s) \qquad \text{pointwise in } s \in \mathbf{R}\setminus\{0\},$$

and take the limit $n \to \infty$. Since (z,v) is a classical solution, we have

$$\int_0^T dt \iint_{\mathbf{R}^N\times\mathbf{R}^N} \frac{\chi_{|x-y|<R}(x,y)}{\omega_N R^N}\chi_\Omega(x)\chi_\Omega(y)\Phi_n'(v_t(x,t))v_t(x,t)z(y,t)dxdy$$

$$\longrightarrow \int_0^T dt \iint_{\mathbf{R}^N\times\mathbf{R}^N} \frac{\chi_{|x-y|<R}(x,y)}{\omega_N R^N}\chi_\Omega(x)\chi_\Omega(y))|v_t(x,t)|z(y,t)dxdy$$

and hence

$$\limsup_{n\to\infty}\int_0^T dt \iint_{\mathbf{R}^N\times\mathbf{R}^N} \frac{\chi_{|x-y|<R}(x,y)}{\omega_N R^N}\chi_\Omega(x)\chi_\Omega(y)\times$$

$$\times \big[\Phi_n'(v_t(x,t))-\Phi_n'(z_t(y,t))\big]\,v_t(x,t)z(y,t)\,dxdy \le 0$$

since $z \leq 0$ and

$$\left[\Phi'_n(v_t(x, t)) - \Phi'_n(z_t(y, t))\right] v_t(x, t) \geq \Phi'_n(v_t(x, t))v_t(x, t) - |v_t(x, t)|.$$

Similarly we have

$$\limsup_{n \to \infty} \int_0^T dt \iint_{\mathbf{R}^N \times \mathbf{R}^N} \frac{\chi_{|x-y|<R}(x, y)}{\omega_N R^N} \chi_\Omega(x)\chi_\Omega(y) \times$$
$$\times \left[\Phi'_n(v_t(x, t)) - \Phi'_n(z_t(y, t))\right] v(x, t)z_t(y, t) \, dxdy \leq 0$$

and hence

$$\int_\Omega \left[|v_t(x, T)| + |z_t(x, T)|\right] dx \leq \int_\Omega \left[|v_{0t}(x)| + |z_{0t}(x)|\right] dx.$$

$\square$

Proof of Theorem 4.2   By Lemmas 4.2.1, 4.2.2 and the compact embedding $W^{1,1}(Q_T) \hookrightarrow L^1(Q_T)$, there is a subsequence $\varepsilon_j \downarrow 0$, denoted by $\varepsilon \downarrow 0$ for simplicity, such that

$$v^\varepsilon \longrightarrow v, \; z^\varepsilon \longrightarrow z \qquad \text{a.e. and strongly in } L^1(Q_T)$$
$$\nabla v^\varepsilon \rightharpoonup \nabla v, \; \nabla z^\varepsilon \rightharpoonup \nabla z \qquad \text{weakly in } L^2(Q_T). \tag{4.75}$$

Since $z^\varepsilon \leq 0 \leq v^\varepsilon$, it also holds that

$$z \leq 0 \leq v \quad \text{a.e. in } Q_T. \tag{4.76}$$

We have, on the other hand,

$$\iint_{Q_T} \left[\frac{z^\varepsilon(x, t)^2}{\omega_N R^N} \int_{B(x,R)\cap\Omega} v^\varepsilon(y, t)dy - \frac{v^\varepsilon(x, t)^2}{\omega_N R^N} \int_{B(x,R)\cap\Omega} z^\varepsilon(y, t)dy\right] dxdt$$
$$\leq C_1 d_0 \varepsilon$$

by Lemma 4.2.1, which implies that

$$\frac{z^2}{\omega_N R^N} \int_{B(\cdot,R)\cap\Omega} v \, dy = \frac{v^2}{\omega_N R^N} \int_{B(\cdot,R)\cap\Omega} z \, dy = 0 \qquad \text{a.e. in } Q_T$$

and hence

$$z(x, t) \cdot v(y, t) = 0 \qquad \text{for a.e. } x, y \in \Omega, \; |x - y| < R, t \geq 0.$$

$\square$

Decay rates similar to (4.61) and (4.63) are also valid for (4.55)–(4.56) although the estimates from below are open (see [63]). In fact, the ODE part to this non-local system is not defined.

Here we prove the convergence, following [62].

Theorem 4.3 *If $\|u_0\|_1 \geq \|v_0\|_1$, then the solution (u, v) to (4.55)–(4.56) satisfies*

$$u(\cdot, t) \longrightarrow u_\infty, \quad v(\cdot, t) \longrightarrow 0 \quad in \ C^m(\bar{\Omega}) \tag{4.77}$$

as $t \uparrow +\infty$, for every $m \in [0, 2)$ with $u_\infty \geq 0$ defined by (4.64).

First we show the total mass conservation.

Lemma 4.2.3 *It holds that*

$$\frac{1}{|\Omega|} \int_\Omega \left(u(x, t) - v(x, t) \right) dx = \bar{u}_0 - \bar{v}_0 = u_\infty. \tag{4.78}$$

Proof First, we have

$$\frac{d}{dt} \int_\Omega (u - v) \, dx = \frac{k}{\omega_N R^N} \int_\Omega \left[u \int_{B(\cdot, R) \cap \Omega} v \, dy - v \int_{B(\cdot, R) \cap \Omega} u \, dy \right] dx.$$

Here the right-hand side is equal to $k/\omega_N R^N$ times

$$\iint_{\mathbf{R}^N \times \mathbf{R}^N} \chi_\Omega(x) \chi_\Omega(y) \chi_{|x-y|<R}(x, y) \times$$
$$\times \left[u(x, t) v(y, t) - u(y, t) v(x, t) \right] dx \, dy = 0$$

by Fubini's theorem and the symmetry of $\chi_{|x-y|<R}(x, y)$. Hence (4.78) follows. $\square$

Below we use several semigroup estimates for the relevant operator in $L^p(\Omega)$, that is,

$$\mathcal{B}_p(w) = (-d_2 \Delta + \alpha) w,$$
$$D(\mathcal{B}_p) = \left\{ w \in W^{2, p}(\Omega) \mid \left. \frac{\partial w}{\partial \nu} \right|_{\partial \Omega} = 0 \right\}$$

where $\alpha > 0$ and $1 < p < \infty$ (see [42, 108]). The spectrum of $\mathcal{B}_p$ lies on the positive axis and a resolvent estimate holds, and hence the fractional powers $\mathcal{B}_p^\gamma$, $0 \leq \gamma < 1$, are defined. The operator $\mathcal{B}_p$ also generates an analytic semigroup denoted by $\{e^{-t\mathcal{B}_p}\}_{t \geq 0}$. There holds

$$\left\| \mathcal{B}_p^\gamma e^{-t\mathcal{B}_p} w \right\|_p \leq C_4(\gamma) q^{-\gamma}(t) e^{-\alpha t} \|w\|_p, \quad t > 0, \tag{4.79}$$

where $0 \leq \gamma < 1$ and $0 < q(t) = \min\{t, 1\} \leq 1$. It is obvious that

$$\int_0^t q^{-\gamma}(\sigma) e^{\delta\sigma} \, d\sigma \leq \begin{cases} C_5(\gamma, \delta) e^{\delta t}, & \text{if } \delta > 0 \\ C_5(\gamma, \delta) (t+1), & \text{if } \delta = 0 \\ C_5(\gamma, \delta), & \text{if } \delta < 0. \end{cases}$$

Finally, given $m \in [0, 2)$ and $p \in (1, +\infty)$ satisfying

$$m < 2\beta - N/p, \quad \beta \in (0, 1),$$

we obtain the embedding

$$D(\mathcal{B}_p^\beta) \subset C^m(\overline{\Omega}) \tag{4.80}$$

(see [42]).

Proof of Theorem 4.3 We convert the second equations of (4.68)–(4.69) to

$$v(\cdot, t) = e^{-t\mathcal{B}_p} v_0 + \alpha \int_0^t e^{-(t-s)\mathcal{B}_p} v(\cdot, s) \, ds$$

$$- \frac{k}{\omega_N R^N} \int_0^t e^{-(t-s)\mathcal{B}_p} \Big[v(\cdot, s) \cdot \int_{B(\cdot, R) \cap \Omega} u(y, s) \, dy \Big] ds, \tag{4.81}$$

which yields

$$\mathcal{B}_p^\gamma v(\cdot, t) = \mathcal{B}_p^\gamma e^{-t\mathcal{B}_p} v_0 + \alpha \int_0^t \mathcal{B}_p^\gamma e^{-(t-s)\mathcal{B}_p} v(\cdot, s) \, ds$$

$$- \frac{k}{\omega_N R^N} \int_0^t \mathcal{B}_p^\gamma e^{-(t-s)\mathcal{B}_p} \Big[v(\cdot, s) \cdot \int_{B(\cdot, R) \cap \Omega} u(y, s) \, dy \Big] ds.$$

Let $t \geq \delta > 0$. Due to (4.79) and (4.58), we have

$$\left\| \mathcal{B}_p^\gamma e^{-t\mathcal{B}_p} v_0 \right\|_p \leq C_6 q^{-\gamma}(t) e^{-\alpha t} \|v_0\|_p \leq C_7 \tag{4.82}$$

and

$$\int_0^t \left\| \mathcal{B}_p^\gamma e^{-(t-s)\mathcal{B}_p} v(\cdot, s) \right\|_p \, ds \leq C_7 \int_0^t q^{-\gamma}(s) e^{-\alpha s} \, ds \leq C_8. \tag{4.83}$$

Finally, we have

$$\int_0^t \left\| \mathcal{B}_p^\gamma e^{-(t-s)\mathcal{B}_p} v(\cdot, s) \frac{1}{\omega_N R^N} \int_{B(\cdot, R) \cap \Omega} u(y, s) \, dy \right\|_p \, ds \leq C_9, \tag{4.84}$$

because (4.58) implies

$$\left\| v(\cdot, t) \cdot \frac{1}{\omega_N R^N} \int_{B(\cdot, R) \cap \Omega} u(y, t) \, dy \right\|_{\infty} \le \|u_0\|_{\infty} \|v_0\|_{\infty}.$$

Combining (4.82), (4.83), and (4.84), we obtain

$$\left\| \mathcal{B}_p^\gamma v(\cdot, t) \right\|_p \le C_{10}, \quad t \ge \delta \tag{4.85}$$

and hence the orbit $\{v(t)\}_{t \ge \delta}$ is compact in $C(\overline{\Omega})$ by Morrey's embedding theorem. Thus there exist $t_j \uparrow +\infty$ and $v^* \in C(\overline{\Omega})$ such that

$$v(\cdot, t_j) \longrightarrow v^* \quad \text{in } C(\overline{\Omega}), \quad j \to \infty. \tag{4.86}$$

Similarly we can show that

$$t \longmapsto \mathcal{B}_p^\gamma v(\cdot, t) \in L^p(\Omega) \quad \text{is Hölder continuous in } [\delta, +\infty), \tag{4.87}$$

by Lemma 3.1 of [51]. Then similarly to Lemma 8 of [85], it follows that

$$t \longmapsto \|\nabla v(\cdot, t)\|_2^2 \quad \text{is uniformly continuous in } [\delta, +\infty). \tag{4.88}$$

Multiplying the second equation in (4.68)–(4.69) by v, we obtain

$$\frac{1}{2} \|v(\cdot, t)\|_2^2 + d_2 \int_0^t \|\nabla v(\cdot, s)\|_2^2 \, ds$$

$$+ \frac{k}{\omega_N R^N} \int_0^t ds \int_\Omega dx \cdot v^2(x, t) \left[\int_{B(x, R) \cap \Omega} u(y, t) \, dy \right] = \frac{1}{2} \|v_0\|_2^2 \tag{4.89}$$

whence

$$\int_0^\infty \|\nabla v(\cdot, s)\|_2^2 \, ds < +\infty. \tag{4.90}$$

An immediate consequence of (4.88) and (4.90) is that

$$\|\nabla v(\cdot, t)\|_2^2 \longrightarrow 0, \quad t \uparrow +\infty. \tag{4.91}$$

Using Poincaré's inequality

$$\mu_2 \|v(\cdot, t) - \overline{v}(\cdot, t)\|_2 \le \|\nabla v(\cdot, t)\|_2$$

as well as (4.86) and (4.91), we see that

$$v(\cdot, t_j) \longrightarrow v^* = \text{constant} \quad \text{in } C(\overline{\Omega}), \quad j \to \infty. \tag{4.92}$$

Analogous properties to (4.85) and (4.86) hold also to u. Therefore, there exists a subsequence of $\{t_j\}$, denoted by the same symbol, and $u^* \in C(\overline{\Omega})$, such that

$$u(\cdot, t_j) \longrightarrow u^* = \text{constant} \quad \text{in } C(\overline{\Omega}), \quad j \longrightarrow \infty. \tag{4.93}$$

By (4.87) and Morrey's embedding theorem, we finally obtain that

$$t \longmapsto \|v(\cdot, t)\|_\infty^2 \quad \text{is uniformly continuous in } [\delta, +\infty),$$

and, similarly,

$$t \longmapsto \|u(\cdot, t)\|_\infty^2 \quad \text{is uniformly continuous in } [\delta, +\infty).$$

Then it holds that

$$t \longmapsto \left\| \frac{v(\cdot, t)^2}{\omega_N R^N} \int_{B(\cdot, R) \cap \Omega} u(y, t) \, dy \right\|_1 \quad \text{is uniformly continuous in } [\delta, +\infty).$$

We also have

$$\frac{v(\cdot, t)^2}{\omega_N R^N} \int_{B(\cdot, R) \cap \Omega} u(y, t) \, dy \in L^1(0, +\infty; L^1(\Omega)),$$

by (4.89), and hence

$$\left\| \frac{v(\cdot, t)^2}{\omega_N R^N} \int_{B(\cdot, R) \cap \Omega} u(y, t) \, dy \right\|_1 \longrightarrow 0, \quad t \uparrow +\infty. \tag{4.94}$$

Here we conclude that

$$\frac{v(\cdot, t_j)^2}{\omega_N R^N} \int_{B(\cdot, R) \cap \Omega} u(y, t_j) \, dy \longrightarrow F(\cdot, R)(v^*)^2 u^* \quad \text{in } L^1(\Omega), \quad j \longrightarrow \infty,$$

by (4.92) and (4.93), where $F(x, R) = |B(x, R) \cap \Omega| / |B(x, R)|$. Then (4.94) implies

$$(v^*)^2 u^* = 0, \tag{4.95}$$

because $F(x, R) \neq 0$. On the other hand,

$$u^* - v^* = u_\infty \tag{4.96}$$

by (4.92), (4.93), and (4.78). If $u_\infty > 0$, equalities (4.95) and (4.96) imply $u^* = u_\infty$, $v^* = 0$, while in the complementary case $u_\infty = 0$ we get $u^* = v^* = 0$. Then the uniqueness of the above limits entails

$$u(\cdot, t) \longrightarrow u_\infty, \quad v(\cdot, t) \longrightarrow 0 \quad \text{in } C(\overline{\Omega}), \quad t \uparrow +\infty.$$

In order to prove the last part of the theorem we use the following interpolation inequality [42]:

$$\|\mathcal{B}_p^\theta v(\cdot, t)\|_p \leq C_{11} \|\mathcal{B}_p^\gamma v(\cdot, t)\|_p^{\theta/\gamma} \|v(\cdot, t)\|_p^{1-\theta/\gamma}, \quad 0 < \theta < \gamma < 1. \tag{4.97}$$

Since $\|\mathcal{B}_p^\gamma v(\cdot, t)\|_p$ is bounded by (4.85), inequality (4.97) yields

$$\|\mathcal{B}_p^\theta v(\cdot, t)\|_p \longrightarrow 0, \quad t \uparrow +\infty, \quad \theta \in (0, 1). \tag{4.98}$$

Then (4.80) implies

$$\|v(\cdot, t)\|_{C^m(\overline{\Omega})} \longrightarrow 0, \quad t \uparrow +\infty, \quad m \in [0, 2).$$

Similarly,
$$\|u(\cdot, t) - u_\infty\|_{C^m(\overline{\Omega})} \longrightarrow 0, \quad t \uparrow +\infty.$$

$\square$

4.3 Reaction-Diffusion Systems

Full system of chemotaxis (4.1) is provided with the other variational structure [123]. First, the total mass conservation arises as (2.54). Next, writing the Smoluchowski part as (2.56), we obtain (2.57). This time the left-hand side is equal to

$$\frac{d}{dt} \int_\Omega [u(\log u - 1) - uv] \, dx + \int_\Omega uv_t \, dx.$$

It holds that

$$\int_\Omega uv_t \, dx = \left\langle \tau v_t - \Delta v + \frac{1}{|\Omega|} \int_\Omega u, v_t \right\rangle = \tau \|v_t\|_2^2 + (\nabla v, \nabla v_t)$$

$$= \tau \|v_t\|_2^2 + \frac{1}{2} \frac{d}{dt} \|\nabla v\|_2^2$$

since $\int_\Omega v = 0$, and we end up with

$$\frac{d}{dt} \int_\Omega \left[u(\log u - 1) + \frac{1}{2}|\nabla v|^2 - uv \right] dx$$

$$= -\tau \|v_t\|^2 - \int_\Omega u |\nabla(\log u - v)|^2 \, dx \le 0. \tag{4.99}$$

Actually this system is a model (B)—model (A) equation associated with the functional

$$L(u, v) = \int_\Omega u(\log u - 1) \, dx + \frac{1}{2}\|\nabla v\|_2^2 - \langle v, u \rangle$$

called Lagrangian in the context of game theory [134, 135], which is defined for

$$0 \le u \in L^1(\Omega), \quad v \in H^1(\Omega), \quad \int_\Omega v = 0. \tag{4.100}$$

That is

$$u_t = \nabla \cdot u \nabla L_u(u, v), \quad \tau v_t = -L_v(u, v) \quad \text{in } \Omega \times (0, T),$$

$$u \frac{\partial}{\partial \nu} L_u(u, v) \bigg|_{\partial\Omega} = 0. \tag{4.101}$$

Relations (2.54) and (4.99) are direct consequences of this formulation, indeed,

$$\frac{d}{dt} \int_\Omega u \, dx = \int_{\partial\Omega} \nu \cdot u \nabla L_u(u, v) \, dx = 0$$

$$\frac{d}{dt} L(u, v) = \int_\Omega \left[L_u(u, v)u_t + L_v(u, v)v_t \right] dx$$

$$= -\int_\Omega \left[u|\nabla L_u(u, v)|^2 + \tau v_t^2 \right] dx \le 0.$$

The stationary state, on the other hand, is formulated equivalently for both components u and v, that is, (2.67) and (2.68), respectively. Both have variational structures. They are Euler-Lagrange equations for

$$\mathcal{F}(u) = \int_\Omega u(\log u - 1) \, dx - \frac{1}{2} \langle (-\Delta)^{-1} u, u \rangle, \quad u \ge 0, \ \|u\|_1 = \lambda$$

and

$$\mathcal{J}_\lambda(v) = \frac{1}{2}\|\nabla v\|_2^2 - \lambda \log \left(\int_\Omega e^v \, dx \right) + \lambda(\log \lambda - 1), \quad \int_\Omega v \, dx = 0,$$

system	consistency	dynamics	ensemble
isolated	energy	entropy	micro-canonical
closed	temperature	Helmoltz free energy	canonical
open	pressure	Gibbs free energy	grand-canonical

particle density   **field potential**

$$\overset{\text{duality}}{\longleftrightarrow}$$

Smoluchowski   **Poisson**

$$u_t = \nabla \cdot (\nabla u - u\nabla v) \text{ in } \Omega \times (0,T)$$

$$\frac{\partial u}{\partial \nu} - u\frac{\partial v}{\partial \nu} = 0 \text{ on } \partial\Omega \times (0,T)$$

$$-\Delta v = u - \frac{1}{|\Omega|}\int_\Omega u, \ \left.\frac{\partial v}{\partial \nu}\right|_{\partial\Omega} = 0$$

$$\int_\Omega v = 0$$

$$\Leftrightarrow v = G * u = \int_\Omega G(\cdot, x')u(x')dx'$$

free energy   **model (B) equation**

$$\mathcal{F}(u) = \int_\Omega u(\log u - 1) - \frac{1}{2}\langle G * u, u\rangle$$

$$\delta\mathcal{F}(u) = \log u - G * u$$

$$u_t = \nabla u \cdot \nabla \delta\mathcal{F}(u), \ \left.\frac{\partial}{\partial \nu}\delta\mathcal{F}(u)\right|_{\partial\Omega} = 0$$

Fig. 4.1 Duality in the Smoluchowski-Poisson equations

respectively, where $\lambda = \|u\|_1$. If $(\overline{u}, \overline{v})$ is the stationary state, that is, if $\overline{u}$ and $\overline{v}$ are the solutions to (2.67) and (2.68), respectively, then it follows that

$$L(u, v) \geq \mathcal{F}(u) - L(u, \overline{v}), \quad L(u, v) \geq \mathcal{J}_\lambda(v) - L(\overline{u}, v) \qquad (4.102)$$

for any (u, v) satisfying (4.100). We call unfolding and minimality the inequalities and equalities of (4.102), respectively (Fig. 4.1). These structures are together called *Toland duality*.

We can find other variational structures in biological models provided with a mesoscopic principle. Namely, we have the Toland, Kuhn-Tucker, and Hamilton structures realized in the reaction-diffusion system

$$u_t = \varepsilon^2 \Delta u + f(u, v), \ \tau v_t = D\Delta v + g(u, v) \quad \text{in } \Omega \times (0, T),$$

$$\frac{\partial u}{\partial \nu} = \frac{\partial v}{\partial \nu} = 0 \qquad\qquad \text{on } \partial\Omega \times (0, T),$$

$$u|_{t=0} = u_0(x) > 0, \ v|_{t=0} = v_0(x) > 0 \qquad \text{in } \Omega. \qquad (4.103)$$

Generation of non-constant stationary solutions and convergence of the non-stationary solutions to them have been observed, while existence of the periodic orbits to the ODE part

$$\frac{d\overline{u}}{dt} = f(\overline{u}, \overline{v}), \quad \overline{u}|_{t=0} = \overline{u}_0 > 0,$$

$$\tau\frac{d\overline{v}}{dt} = g(\overline{u}, \overline{v}), \quad \overline{v}|_{t=0} = \overline{v}_0 > 0 \tag{4.104}$$

are known in several cases. The shadow system

$$U_t = \varepsilon^2 \Delta U + f(U, V) \qquad \text{in } \Omega \times (0, T),$$

$$\tau\frac{dV}{dt} = \frac{1}{|\Omega|} \int_\Omega g(U, V)dx \qquad t \in (0, T),$$

$$\frac{\partial U}{\partial \nu} = 0 \qquad\qquad\qquad \text{on } \partial\Omega \times (0, T),$$

$$U|_{t=0} = u_0(x) > 0 \qquad\qquad \text{in } \Omega,$$

$$V|_{t=0} = \overline{v}_0 > 0, \tag{4.105}$$

also arises in the limit $D \uparrow +\infty$. It can happen that the approximation of (4.103) is valid in the transient range $t \sim 1$, while (4.105) is close to (4.104) asymptotically, $t \gg 1$.

First, a mass conserved reaction-diffusion system is associated with the Toland duality. It takes the form

$$u_t = D\Delta u + f(u, v), \ \tau v_t = \Delta v - f(u, v) \quad \text{in } \Omega \times (0, T) \tag{4.106}$$

subject to the boundary condition

$$\frac{\partial u}{\partial \nu} = \frac{\partial v}{\partial \nu} = 0 \quad \text{on } \partial\Omega \times (0, T) \tag{4.107}$$

which realizes the total mass conservation

$$\frac{d}{dt} \int_\Omega (u + \tau v) \ dx = 0. \tag{4.108}$$

The work [102] introduces with a top-down modeling a system of molecular dynamics inside cell (*cell polarity*) associated with chemotaxis. Thus one considers three kind of molecules, RAC, $Cdc42$, and $RhoA$ that are interacting. Each of them has two phases, active and inactive, which are characterized by slow and fast diffusion, respectively (Fig. 4.2). Model (4.106)–(4.107) is focused on these two phases of one

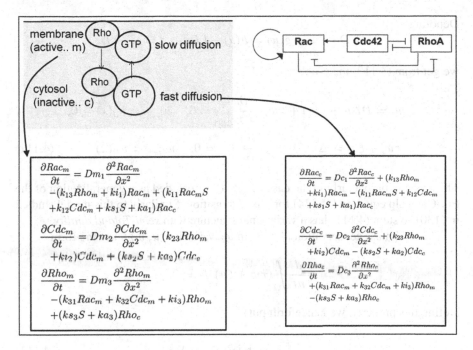

Fig. 4.2 Cell polarization

species, ignoring interactions between the other species. According to [55, 102], a Turing pattern is expected to arise, using the cases

$$f(u, v) = -\frac{au}{u^2 + b} + v,$$

$$f(u, v) = -a_1 \left[\frac{u + v}{\{a_2(u + v) + 1\}^2} - v \right],$$

$$f(u, v) = a_1(u + v)[(\alpha u + v)(u + v) - a_2]. \tag{4.109}$$

The works [57, 86, 87] studied bifurcation from constant solutions and also global-in-time dynamics.

We consider a generalization of the first case of (4.109), namely,

$$f(u, v) = h(u) + kv,$$

where $h = h(u)$ is a smooth function and k is a positive constant. Then

$$u_t = D\Delta u + h(u) + kv, \quad \left.\frac{\partial u}{\partial \nu}\right|_{\partial \Omega} = 0, \quad u|_{t=0} = u_0(x),$$

$$\tau v_t = \Delta v - h(u) - kv, \quad \left.\frac{\partial v}{\partial \nu}\right|_{\partial \Omega} = 0, \quad v|_{t=0} = v_0(x). \tag{4.110}$$

Denoting
$$w = Du + v, \quad g(u) = h(u) - kDu, \quad \xi = 1 - \tau D,$$

we get from (4.110) the problem

$$u_t = D\Delta u + g(u) + kw, \quad \left.\frac{\partial u}{\partial \nu}\right|_{\partial\Omega} = 0, \quad u|_{t=0} = u_0(x) \geq 0,$$

$$\tau w_t + \xi u_t = \Delta w, \quad \left.\frac{\partial w}{\partial \nu}\right|_{\partial\Omega} = 0, \quad w|_{t=0} = w_0(x). \tag{4.111}$$

This system, particularly in the case of $\xi > 0$, is regarded as a generalization of the Fix-Caginalp equation [15, 34] for phase transition. Similarly to this model studied in [130], system (4.111) has a variational structure with *semi-Toland duality*.

The total mass conservation (4.108) in the variables (u, w) takes the form

$$\frac{d}{dt} \int_\Omega (\tau w + \xi u) \, dx = 0.$$

Noting this property, we henceforth put

$$\int_\Omega (\tau w + \xi u) \, dx = \lambda. \tag{4.112}$$

The system admits also a Lyapunov function. In fact, multiplying the first and second equations u_t and w, respectively, we have

$$\|u_t\|_2^2 + \frac{d}{dt} \int_\Omega \frac{D}{2}|\nabla u|^2 - G(u) \, dx = k(w, u_t),$$

and

$$\frac{\tau}{2} \frac{d}{dt}\|w\|_2^2 + \|\nabla w\|_2^2 + \xi(u_t, w) = 0,$$

where $G(u) = \int_0^u g(u)du$ and $(\,,\,)$ is the L^2-inner product. Then it follows that

$$\xi\|u_t\|_2^2 + \xi\frac{d}{dt} \int_\Omega \left(\frac{D}{2}|\nabla u|^2 - G(u)\right) dx + \frac{\tau k}{2} \frac{d}{dt}\|w\|_2^2 + k\|\nabla w\|_2^2 = 0,$$

which means that
$$\frac{d}{dt} L(u, w) = -\|u_t\|_2^2 - \frac{k}{\xi}\|\nabla w\|_2^2 \leq 0, \tag{4.113}$$

where
$$L(u, w) = \int_\Omega \left(\frac{D}{2}|\nabla u|^2 - G(u) + \frac{\tau k}{2\xi}w^2\right) dx.$$

By the theory of dynamical systems and parabolic regularity, if the solution is global-in-time and satisfies $\sup_{t \geq 0} \|u(\cdot, t), v(\cdot, t)\|_\infty < +\infty$, then the ω-limit set, defined by

$$\omega(u_0, v_0) = \{(u_*, v_*) \in C^2(\overline{\Omega})^2 \mid \exists t_k \uparrow +\infty,$$
$$\lim_{k \to \infty} \|u(\cdot, t_k) - u_*, v(\cdot, t_k) - v_*\|_{C^2} = 0\},$$

is non-empty, compact, and connected. From (4.113) it follows that each $(u, v) \in \omega(u_0, v_0)$ satisfies $u_t = 0$ and $w = \overline{w} \in \mathbf{R}$. Since this $\overline{w}$ is determined by (4.112), we have

$$\overline{w} = \frac{1}{\tau |\Omega|} \left(\lambda - \xi \int_\Omega u \, dx \right).$$

Consequently,

$$- D\Delta u = g(u) + \frac{k}{\tau |\Omega|} \left(\lambda - \xi \int_\Omega u \, dx \right), \quad \frac{\partial u}{\partial \nu} \bigg|_\Omega = 0, \qquad (4.114)$$

which is the Euler-Lagrange equation associated with the variational functional

$$J_\lambda(u) = \int_\Omega \left(\frac{D}{2} |\nabla u|^2 - G(u) - \frac{k\lambda}{\tau |\Omega|} u \right) dx + \frac{k\xi}{2\tau |\Omega|} \left(\int_\Omega u \, dx \right)^2,$$

defined for $u \in H^1(\Omega)$.

A remarkable structure is the semi-duality. First, we have the unfolding

$$L|_{w = \frac{1}{\tau |\Omega|} (\lambda - \xi \int_\Omega u \, dx)} = \int_\Omega \left(\frac{D}{2} |\nabla u|^2 - G(u) \right) dx + \frac{k}{2\xi |\Omega| \tau} \left(\lambda - \xi \int_\Omega u \right)^2$$
$$= J_\lambda(u) + \frac{k\lambda^2}{2\xi |\Omega| \tau}.$$

Minimality also holds, which means that if $w = w(x)$ satisfies (4.112), then

$$L(u, w) \geq L(u, \overline{w}), \quad \overline{w} = \frac{1}{|\Omega| \tau} \left(\lambda - \xi \int_\Omega u \, dx \right).$$

This property is actually a consequence of the Schwarz inequality.

When the nonlinearity $G(u)$ is real-analytic the general theory says that any local minimum of J_λ, denoted by $u^* = u^*(x) \in H^1(\Omega)$, is dynamically stable [103]. More precisely, for any $\varepsilon > 0$ there exists a $\delta > 0$ such that

$$\|u_0 - u^*, v_0 - v^*\|_{H^1} < \delta, \quad \int_\Omega (u_0 + \tau v_0) \, dx = \lambda$$

imply

$$\sup_{t\in[0,T)} \|u(\cdot, t) - u^*\|_{H^1} < \varepsilon.$$

This property implies also the stability of w and hence that of v-component [78].
   In the original model, the nonlinearity

$$h(u) = -\frac{au}{u^2 + b}$$

is smooth and decays at $u = +\infty$ together with its derivatives. Therefore, we have
the global-in-time existence and compactness of the orbit by using the Lyapunov
function $L(u, w)$.
   Now we turn to the *Kuhn-Tucker duality*. A skew-gradient system

$$u_t = d_u \Delta u + H_u(u, v), \quad \left.\frac{\partial u}{\partial \nu}\right|_{\partial\Omega} = 0,$$

$$\tau v_t = d_v \Delta v - H_v(u, v), \quad \left.\frac{\partial v}{\partial \nu}\right|_{\partial\Omega} = 0, \qquad (4.115)$$

is realized in Gierer-Meinhardt and FitzHugh-Nagumo models [148, 149]. Using the
skew Lagrangian

$$L(u, v) = \int_\Omega \left(\frac{d_u}{2}|\nabla u|^2 - \frac{d_v}{2}|\nabla v|^2 - H(u, v)\right) dx, \quad u, v \in H^1(\Omega),$$

we have

$$u_t = -L_u(u, v), \quad \tau v_t = L_v(u, v)$$

by (4.115). Then we call $(\overline{u}, \overline{v}) \in H^1(\Omega) \times H^1(\Omega)$ a *saddle* if

$$L(u, \overline{v}) \geq L(\overline{u}, \overline{v}) \geq L(\overline{u}, v), \quad \forall(u, v) \in U \times V \qquad (4.116)$$

where $U \times V$ is an open set in $H^1(\Omega) \times H^1(\Omega)$ containing $(\overline{u}, \overline{v})$. This relation
arises in the context of Kuhn-Tucker duality (see [123], for example). The linearized
system of (4.115) is given by

$$\frac{d}{dt}\begin{pmatrix} u \\ v \end{pmatrix} + JA\begin{pmatrix} u \\ v \end{pmatrix} = 0,$$

where

$$J = \begin{pmatrix} 1 & 0 \\ 0 & \tau^{-1} \end{pmatrix}$$

and

$$A = \begin{pmatrix} -d_u\Delta - H_{uu}(\overline{u}, \overline{v}) & -H_{uv}(\overline{u}, \overline{v}) \\ H_{uv}(\overline{u}, \overline{v}) & -d_v\Delta + H_{vv}(\overline{u}, \overline{v}) \end{pmatrix},$$

$$D(A) = \left\{ (u, v) \in H^2(\Omega) \times H^2(\Omega) \mid \frac{\partial}{\partial\nu}(u, v)\Big|_{\partial\Omega} = 0 \right\}.$$

Its linearized stability means the positivity of the real part of any eigenvalue μ of the eigenvalue problem

$$(\mu I + JA)\begin{pmatrix} u \\ v \end{pmatrix} = 0, \quad \begin{pmatrix} u \\ v \end{pmatrix} \in D(A) \setminus \{0\},$$

which in turn is equivalent to

$$\text{Re }(JAw, w)_J = \text{Re }(Aw, w) > 0, \quad \forall w - \begin{pmatrix} u \\ v \end{pmatrix} \neq 0,$$

where $(z, w)_J = (J^{-1/2}z, J^{-1/2}w)$ and

$$(z, w) = \int_\Omega (z_1\overline{w}_1 + z_2\overline{w}_2)\, dx, \quad z = \begin{pmatrix} z_1 \\ z_2 \end{pmatrix}, \quad w = \begin{pmatrix} w_1 \\ w_2 \end{pmatrix}.$$

Since

$$\text{Re }(Aw, w) = (-d_u\Delta u - H_{uu}(\overline{u}, \overline{v})u, u) + (-d_v\Delta v + H_{vv}(\overline{u}, \overline{v})v, v),$$

this property is equivalent to the positivity of the self-adjoint operators in (real) $L^2(\Omega)$,

$$A_1 = -d_u\Delta - H_{uu}(\overline{u}, \overline{v}), \quad D(A_1) = \left\{ u \in H^2(\Omega) \mid \frac{\partial u}{\partial\nu}\Big|_{\partial\Omega} = 0 \right\},$$

$$A_2 = -d_v\Delta + H_{vv}(\overline{u}, \overline{v}), \quad D(A_2) = \left\{ v \in H^2(\Omega) \mid \frac{\partial v}{\partial\nu}\Big|_{\partial\Omega} = 0 \right\}.$$

Therefore, the linearized stability of the stationary solution $(\overline{u}, \overline{v})$ to (4.115) is a stronger condition than (4.116), that is, $A_i > 0$, $i = 1, 2$.

The Gierer-Meinhardt model arises in morphogenesis [38]. It is a combination of the model (A) equations

$$ra_t = -L_a, \quad q\tau h_t = L_h$$

associated with the skew-Lagrangian

$$L(a, h) = \int_\Omega \left[\frac{r\varepsilon^2}{2} |\nabla a|^2 - \frac{qD}{2} |\nabla h|^2 - H(a, h) \right] dx, \quad a, h \in H^1(\Omega)$$

$$H(a, h) = -\frac{r}{2} a^2 + r\sigma a + a^{p+1} h^{-q} + \frac{q}{2} h^2,$$

that is,

$$a_t = \varepsilon^2 \Delta a - a + \frac{a^p}{h^q} + \sigma, \qquad \left. \frac{\partial a}{\partial \nu} \right|_{\partial \Omega} = 0,$$

$$\tau h_t = D\Delta h - h + \frac{a^r}{h^s}, \qquad \left. \frac{\partial h}{\partial \nu} \right|_{\partial \Omega} = 0. \tag{4.117}$$

Here $r, q, \tau, \varepsilon, D, p > 0, \sigma \geq 0$ are constants and

$$p + 1 = r, \quad q + 1 = s$$

is assumed for (4.117) to reduce to (4.115).

The shadow system arises formally when $D \uparrow +\infty$. More precisely, in this case $h = h(t)$ becomes independent of x. Then we have

$$a_t = \varepsilon^2 \Delta a - a + a^p / h^q, \qquad \left. \frac{\partial a}{\partial \nu} \right|_{\partial \Omega} = 0,$$

$$\tau h_t = -h + \frac{1}{h^s} \cdot \frac{1}{|\Omega|} \int_\Omega a^r \, dx \tag{4.118}$$

by operating $\frac{1}{|\Omega|} \int_\Omega \cdot$ in the second equation, where it is assumed that $\sigma = 0$. Its stationary state is defined by

$$\varepsilon^2 \Delta a - a + \frac{a^p}{h^q} = 0, \quad h^{s+1} = \frac{1}{|\Omega|} \int_\Omega a^r \, dx, \quad \left. \frac{\partial a}{\partial \nu} \right|_{\partial \Omega} = 0,$$

(see [65, 96]). In the case of $r = p + 1$ one has the variational functional

$$J(v) = \frac{1}{2} \int_\Omega \left(\varepsilon^2 |\nabla a|^2 + a^2 \right) dx - \frac{1}{(1 - \gamma)r} \left(\int_\Omega a^r \, dx \right)^{1-\gamma}$$

defined for $v \in H^1(\Omega)$, where $\gamma = \frac{r}{s+1}$. The system (4.117) displays spiky stationary solutions, slow dynamics of spikes, and Hopf bifurcation (see [24, 140, 141]).

The FitzHugh-Nagumo equation describes propagation of nerve impulses [33, 93]. It is a combination of the model (A) equations

$$u_t = -L_u(u, v), \quad \tau v_t = L_v(u, v)$$

derived from the skew Lagrangian

$$L(v, u) = \int_\Omega \left[\frac{\xi^2}{2} |\nabla u|^2 + W(u) - \frac{\sigma}{2} |\nabla v|^2 + uv \right] dx$$

$$W(u) = \frac{1}{4} \left(u^2 - 1 \right)^2, \quad u, v \in H^1(\Omega), \quad \int_\Omega v \, dx = 0,$$

where $\tau, \xi, \sigma > 0$ are constants. The corresponding system is

$$u_t = \xi^2 \Delta u - W'(u) - v, \quad \left. \frac{\partial u}{\partial \nu} \right|_{\partial \Omega} = 0,$$

$$\tau v_t = \sigma \Delta v + u - \frac{1}{|\Omega|} \int_\Omega u \, dx, \quad \left. \frac{\partial v}{\partial \nu} \right|_{\partial \Omega} = 0, \quad \int_\Omega v \, dx = 0. \quad (4.119)$$

A Kuhn-Tucker semi-duality holds and we have

$$\inf_v L = L|_{v = (-\sigma \Delta)^{-1} u} = J^*.$$

Here

$$J^*(u) = \int_\Omega \left(\frac{\xi^2}{2} |\nabla u|^2 + W(u) \right) dx + \frac{\sigma}{2} \langle (-\Delta)^{-1} u, u \rangle$$

stands for the Ohta-Kawasaki free energy [97] and $v = (-\Delta)^{-1} u$ means (2.55) (see [123] for more details).

The final structure we mention here is the *Hamilton formalism*. First, the prey-predator system is a form of Lotka-Volterra model associated with

$$f(u, v) = u(a - bv), \quad g(u, v) = v(-c + du),$$

where $a, b, c, d > 0$ are constants. In this case the ODE part (4.104) has a unique equilibrium $u_* = c/d$, $v_* = a/b$, and the other orbits are periodic, since one has the first integral

$$a \log \overline{v} - b\overline{v} + c \log \overline{u} - d\overline{u} = \text{constant}.$$

This ODE part recasts as a Hamilton system using

$$\xi = \log u, \quad \eta = \log v,$$
$$H(\xi, \eta) = -a\eta + be^\eta - \tau^{-1} c\xi + \tau^{-1} de^\xi, \quad (4.120)$$

that is

$$\xi_t = -H_\eta, \quad \eta_t = H_\xi.$$

Using the variables (4.120), the reaction-diffusion system (4.103) is transformed into

$$\xi_t = \varepsilon^2 e^{-\xi}\Delta e^\xi - H_\eta, \qquad \left.\frac{\partial \xi}{\partial \nu}\right|_{\partial\Omega} = 0,$$

$$\eta_t = \tau^{-1} D e^{-\eta}\Delta e^\eta + H_\xi, \qquad \left.\frac{\partial \eta}{\partial \nu}\right|_{\partial\Omega} = 0,$$

which guarantees that

$$\frac{d}{dt}\int_\Omega H(\xi,\eta)dx = -\tau^{-1}\int_\Omega c\varepsilon^2|\nabla\xi|^2 + aD|\nabla\eta|^2 \, dx \le 0. \tag{4.121}$$

This estimate implies

$$\int_\Omega \left(e^\xi + e^\eta\right) \, dx \le C$$

and hence

$$\limsup_{t\uparrow T}\{\|u(\cdot,t)\|_1 + \|v(\cdot,t)\|_1\} < +\infty.$$

Since

$$u_t \le \varepsilon^2 \Delta u + au, \; u > 0, \qquad \left.\frac{\partial u}{\partial \nu}\right|_{\partial\Omega} = 0,$$

it follows that

$$\limsup_{t\uparrow T}\|u(t)\|_\infty < +\infty. \tag{4.122}$$

Using the equation

$$\tau v_t = D\Delta v + v(-c + du), \; v > 0, \qquad \left.\frac{\partial v}{\partial \nu}\right|_{\partial\Omega} = 0,$$

we obtain

$$\limsup_{t\uparrow T}\|v(\cdot,t)\|_\infty < +\infty \tag{4.123}$$

and then $T = +\infty$ follows together with the pre-compactness the orbit $\{(u(t), v(t)\}_{t\ge 0}$ in $C_+(\overline{\Omega})^2$. Hence the ω-limit set of this orbit must be contained in the set of spatially homogeneous solutions, thanks to the existence of the Lyapunov function (4.121) which also guarantees that the stationary state must be constant.

Theorem 4.4 ([79]) *We have $T = +\infty$ in the Lotka-Volterra system (4.103), together with (4.122)–(4.123). There is an ODE orbit $\mathcal{O} \subset \mathbf{R}^2$ such that*

$$\lim_{t\uparrow+\infty} dist_{C^2}(u(\cdot,t), v(\cdot,t), \mathcal{O}) - 0.$$

If this $\mathcal{O}$ does not reduce to a single point, then there is $\ell > 0$ such that

$$\lim_{t \uparrow +\infty} \|u(\cdot, t + \ell) - u(\cdot, t), v(\cdot, t + \ell) - v(\cdot, t)\|_{C^2} = 0.$$

The above result is more or less known [3, 25]. Inequality (4.122), however, is independent of D. Then the energy method applied to the second equation of (4.103) implies the convergence of v to a function V in $C(\overline{\Omega} \times [0, T])$ which satisfies

$$\tau \frac{dV}{dt} = \frac{V}{|\Omega|} \int_\Omega (-c + du)\, dx, \quad V|_{t=0} = \overline{v}_0 = \frac{1}{|\Omega|} \int_\Omega v_0\, dx$$

as $D \uparrow +\infty$. Then we obtain the shadow system (4.105) with the convergence $u \to U$ in $C(\overline{\Omega} \times [0, T])$. Here, the behavior of the solution (U, V) to (4.105) as $t \uparrow +\infty$ is closer to that described by the ODE system (4.104). In fact,

$$\overline{u} = \frac{1}{|\Omega|} \int_\Omega U\, dx, \quad \overline{v} = V$$

solves (4.104), and, therefore, unless $(\overline{u}_0, \overline{v}_0) = (u_*, v_*)$, it is periodic in time with the period $T > 0$ satisfying

$$\frac{1}{T} \int_0^T \overline{u}(t)\, dt = \frac{c}{d}, \quad \frac{1}{T} \int_0^T \overline{v}(t)\, dt = \frac{a}{b}.$$

To show the convergence of (4.105) to (4.104) as $t \uparrow +\infty$, we put

$$w = U - \overline{u}, \quad g = u - bV,$$

which satisfy

$$w_t = \varepsilon^2 \Delta w + g(t)w, \quad \left.\frac{\partial w}{\partial \nu}\right|_{\partial \Omega} = 0, \quad w|_{t=0} = u_0 - \overline{u}_0 \equiv w_0.$$

It holds also that

$$g(t + T) = g(t), \quad \int_0^T g(t)\, dt = 0, \tag{4.124}$$

while

$$W(x, t) = w(x, t) \exp\left(-\int_0^t g(s)\, ds\right),$$

solves

$$W_t = \varepsilon^2 \Delta W, \quad \left.\frac{\partial W}{\partial \nu}\right|_{\partial \Omega} = 0, \quad W|_{t=0} = w_0.$$

We thus obtain

$$w(\cdot, t) = e^{t\varepsilon^2 \Delta_N} \overline{w_0} \cdot G(t), \quad G(t) = \exp\left(\int_0^t g(s)ds\right)$$

with $G(t + T) = G(t)$, thanks to (4.124). This precise expression, in particular, implies

$$\lim_{t\uparrow+\infty} \|U(\cdot, t) - \overline{u}(t)\|_\infty = 0.$$

A structure similar to that of the Lotka-Volterra ODE is observed for the Gierer-Meinhardt ODE

$$\frac{du}{dt} = -u + \frac{u^p}{v^q}, \quad \tau\frac{dv}{dt} = -v + \frac{u^r}{v^s}, \tag{4.125}$$

with

$$\tau = \frac{s+1}{p-1}. \tag{4.126}$$

In fact, writing (4.125) as

$$u^{-p}(u_t + u) = v^{-q}, \quad v^s(v_t + \tau^{-1}v) = \tau^{-1}u^r,$$

we introduce the new variables,

$$\xi = \frac{u^{-p+1}}{p-1}, \quad \eta = \frac{v^{s+1}}{s+1}$$

which satisfy $\xi_t = -u_t u^{-p}$, $\eta_t = v^s v_t$. Since

$$\xi_t = u^{-p+1} - v^{-q} = (p-1)\xi - \{(s+1)\eta\}^{-\frac{q}{s+1}},$$
$$\eta_t = -\tau^{-1}v^{s+1} + \tau^{-1}u^r = -\tau^{-1}(s+1)\eta + \tau^{-1}\{(p-1)\xi\}^{-\frac{r}{p-1}},$$

we end up with

$$\frac{d\xi}{dt} = H_\eta, \quad \frac{d\eta}{dt} = -H_\xi,$$

where

$$H(\xi, \eta) = (p-1)\xi\eta + \left(\frac{r}{p-1} - 1\right)^{-1} A(\xi) + \left(\frac{q}{s+1} - 1\right)^{-1} B(\eta),$$
$$A(\xi) = \tau^{-1}(p-1)^{-\frac{r}{p-1}} \xi^{1-\frac{r}{p-1}}, \quad B(\eta) = (s+1)^{-\frac{q}{s+1}} \eta^{1-\frac{q}{s+1}},$$

assuming (4.126). Then result analogous to Theorem 4.4 holds true [60].

There is a Hamilton structure even for more than 3 components. Here, we just mention the system

$$\frac{du_1}{dt} = (u_2 - u_3)u_1, \quad \frac{du_2}{dt} = (u_3 - u_1)u_2, \quad \frac{du_3}{dt} = (u_1 - u_2)u_3.$$

Using $\xi_i = \log u_i$, $i = 1, 2, 3$, we have

$$\frac{d\xi_1}{dt} = e^{\xi_2} - e^{\xi_3}, \quad \frac{d\xi_2}{dt} = e^{\xi_3} - e^{\xi_1}, \quad \frac{d\xi_3}{dt} = e^{\xi_1} - e^{\xi_2},$$

which is written as

$$\frac{d\xi}{dt} = H(\xi) \times a \tag{4.127}$$

for

$$\xi = \begin{pmatrix} \xi_1 \\ \xi_2 \\ \xi_3 \end{pmatrix}, \quad a = \begin{pmatrix} 1 \\ 1 \\ 1 \end{pmatrix}, \quad H(\xi) = \begin{pmatrix} e^{\xi_1} \\ e^{\xi_2} \\ e^{\xi_3} \end{pmatrix}.$$

Equation (4.127) implies

$$0 = \frac{d\xi}{dt} \cdot a = \frac{d}{dt}(\xi_1 + \xi_2 + \xi_3) = \frac{d}{dt}a \cdot \xi$$

$$0 = \frac{d\xi}{dt} \cdot H(\xi) = \frac{d}{dt}(e^{\xi_1} + e^{\xi_2} + e^{\xi_3}) = \frac{d}{dt}a \cdot H(\xi).$$

Using this ODE structure, we can prove the following theorem concerning the system

$$\frac{\partial u_1}{\partial t} = d_1 \Delta u_1 + (u_2 - u_3)u_1, \quad \left. \frac{\partial u_1}{\partial \nu} \right|_{\partial \Omega} = 0,$$

$$\frac{\partial u_2}{\partial t} = d_2 \Delta u_2 + (u_3 - u_1)u_2, \quad \left. \frac{\partial u_2}{\partial \nu} \right|_{\partial \Omega} = 0,$$

$$\frac{\partial u_3}{\partial t} = d_3 \Delta u_3 + (u_1 - u_2)u_3, \quad \left. \frac{\partial u_3}{\partial \nu} \right|_{\partial \Omega} = 0,$$

recalling that the space dimension is N.

Theorem 4.5 ([132]) *If $N \le 2$, then it holds that*

$$T = +\infty, \quad \sup_{t \ge 0} \|u(\cdot, t)\|_\infty < +\infty,$$

where $u = {}^t(u_1, u_2, u_3)$. There is an ODE orbit $\mathcal{O}$ such that

$$\lim_{t \uparrow +\infty} dist_C(u(\cdot, t), \mathcal{O}) = 0$$

and if $\sharp \mathcal{O} \neq 1$, then there is $\ell > 0$ such that

$$\lim_{t\uparrow+\infty} \|u(\cdot, t+\ell) - u(\cdot, t)\|_\infty = 0.$$

The above structure and results are valid also for to the general system with anti-symmetric interacting part,

$$\tau_j \frac{\partial u_j}{\partial t} = d_j \Delta u_j + \left(-e_j + \sum_k a_{jk} u_k\right) u_j \quad \text{in } \Omega \times (0, T)$$

$$\left. \frac{\partial u_j}{\partial \nu}\right|_{\partial\Omega} = 0, \quad u_j\big|_{j=0} = u_{j0} \geq 0,$$

where $^t A + A \leq 0$ and $e_j \geq 0$ for all j, $A = (a_{jk})$ (see [133]).

4.4 Competitive System of Chemotaxis

A competitive feature of chemotaxis is observed in tumor-associated microenvironment at the stage of intravasation (Fig. 4.3). More precisely, there is an interaction between cancer cells and macrophages through chemical substances which causes their localized cell deformations, called invadopodia and podosomes, respectively [58, 147]. This is modeled by a competitive chemotactic feature between two species of cells to chemical substance secreted by themselves. Then there is a case that only one component of the solution can form δ-functions at the blowup time [29]. This model is also used for cell sorting of Dd (cellular slime molds) in the mound-stage (see [142] and also [21, 30, 46, 47, 146] for related mathematical studies).

First, we have a parameter region which ensures simultaneous blowup also for non-radially symmetric solutions. If the existence time of the solution is finite, there is a formation of collapse (possibly degenerate) with collapse mass curve. For radially symmetric solutions the collapse concentrates mass on one component if the total masses of the other components are relatively small.

Given a bounded domain $\Omega \subset \mathbf{R}^2$ with smooth boundary $\partial\Omega$, this model combines the part of Smoluchowski equations

$$\partial_t u_1 = d_1 \Delta u_1 - \chi_1 \nabla \cdot u_1 \nabla v,$$
$$\partial_t u_2 = d_2 \Delta u_2 - \chi_2 \nabla \cdot u_2 \nabla v \quad \text{in } \Omega \times (0, T), \tag{4.128}$$

using the chemical gradient term ∇v, with the boundary condition

$$d_1 \frac{\partial u_1}{\partial \nu} - \chi_1 u_1 \frac{\partial v}{\partial \nu} = d_2 \frac{\partial u_2}{\partial \nu} - \chi_2 u_2 \frac{\partial v}{\partial \nu} = 0 \quad \text{on } \partial\Omega \times (0, T) \tag{4.129}$$

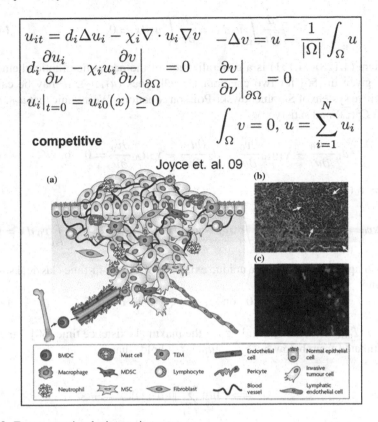

$$u_{it} = d_i \Delta u_i - \chi_i \nabla \cdot u_i \nabla v \quad -\Delta v = u - \frac{1}{|\Omega|} \int_\Omega u$$

$$d_i \frac{\partial u_i}{\partial \nu} - \chi_i u_i \frac{\partial v}{\partial \nu}\bigg|_{\partial\Omega} = 0 \qquad \frac{\partial v}{\partial \nu}\bigg|_{\partial\Omega} = 0$$

$$u_i|_{t=0} = u_{i0}(x) \geq 0$$

competitive $\qquad \int_\Omega v = 0, \; u = \sum_{i=1}^{N} u_i$

Joyce et. al. 09

(a)   (b)   (c)

BMDC	Mast cell	TEM	Endothelial cell	Normal epithelial cell
Macrophage	MDSC	Lymphocyte	Pericyte	Invasive tumour cell
Neutrophil	MSC	Fibroblast	Blood vessel	Lymphatic endothelial cell

Fig. 4.3 Tumor-associated microenvironment

and the initial condition

$$u_1|_{t=0} = u_{10}(x) \geq 0, \quad u_2|_{t=0} = u_{20}(x) \geq 0 \quad \text{in } \Omega, \tag{4.130}$$

where $d_1, d_2, \chi_1,$ and χ_2 are positive constants and ν is the unit normal vector, with the Poisson part normalized so that

$$-\Delta v = u - \frac{1}{|\Omega|} \int_\Omega u \, dx, \quad \frac{\partial v}{\partial \nu}\bigg|_{\partial\Omega} = 0, \quad \int_\Omega v \, dx = 0, \quad u = u_1 + u_2. \tag{4.131}$$

We put

$$\int_\Omega v \, dx = 0$$

in (4.131) to normalize an additive constant of v. This normalization is not essential because only the gradient ∇v of v appears in the Smoluchowski equations (4.128). It holds also that

$$\frac{d}{dt} \int_\Omega u_1 dx = \frac{d}{dt} \int_\Omega u_2 dx = 0. \tag{4.132}$$

System (4.128)–(4.131) is a generalization of the parabolic-elliptic chemotaxis system given in [56] for two chemotactic cell types (u_1, u_2). It may be called a competitive system of Smoluchowski-Poisson equations. We assume that $u_{10}, u_{20} \neq 0$ are in $C^2(\overline{\Omega})$ and satisfy

$$d_1 \frac{\partial u_{10}}{\partial \nu} - \chi_1 u_{10} \frac{\partial v_0}{\partial \nu} = d_2 \frac{\partial u_{20}}{\partial \nu} - \chi_2 u_{20} \frac{\partial v_0}{\partial \nu} = 0 \quad \text{on } \partial\Omega$$

for $v_0 = v_0(x)$ defined by

$$-\Delta v_0 = u_{10} + u_{20} - \frac{1}{|\Omega|} \int_\Omega u_{10} + u_{20} \, dx, \quad \left.\frac{\partial v_0}{\partial \nu}\right|_{\partial\Omega} = 0, \quad \int_\Omega v_0 \, dx = 0.$$

This assumption guarantees the unique existence of a local-in-time classical solution satisfying

$$u_i(\cdot, t) > 0 \quad \text{on } \overline{\Omega}, \quad 0 < t < T, \, i = 1, 2, \tag{4.133}$$

with $T = T_{\max} \in (0, +\infty]$ standing for the maximal existence time [30]. Since this T is estimated from below by $\sum_{i=1}^2 \|u_{i0}\|_\infty$, it holds that

$$T < +\infty \implies \lim_{t\uparrow T} \sum_{i=1}^2 \|u_i(\cdot, t)\|_\infty = +\infty, \tag{4.134}$$

by a standard argument on the continuation of the solution in time (see, for example, [128]).

In two space dimensions with $u_i = u_i(|x|, t)$, the simultaneous blowup

$$T < +\infty \implies \limsup_{t\uparrow T} \|u_1(\cdot, t)\|_\infty = \limsup_{t\uparrow T} \|u_2(\cdot, t)\|_\infty = +\infty, \tag{4.135}$$

takes place for any pairs of (d_i, χ_i), $i = 1, 2$ (see [30, 31, 71] and [71] for the proof and related results). Here we review several results obtained in [29]. The first theorem displays a parameter region where the simulatenous blowup occurs even for non-radially symmetric solutions. Let

$$\xi_i = d_i/\chi_i, \quad \|u_{i0}\|_1 = \lambda_i, \quad i = 1, 2, \tag{4.136}$$

be the inverse motilities and initial masses of the species.

Theorem 4.6 *If*

$$\lambda_i < 4\pi\xi_i, \quad i = 1, 2, \tag{4.137}$$

then it holds that

$$T < +\infty \implies \lim_{t\uparrow T} \|u_1(\cdot, t)\|_\infty = \lim_{t\uparrow T} \|u_2(\cdot, t)\|_\infty = +\infty. \tag{4.138}$$

We note that condition (4.137) is consistent with the assumption $T < +\infty$. In fact, if

$$\left(\sum_{i=1}^2 \lambda_i\right)^2 < 4\pi \sum_{i=1}^2 \xi_i\lambda_i, \quad \lambda_i < 4\pi\xi_i, \; i = 1, 2,$$

then it always holds that $T = +\infty$, while $T < +\infty$ can occur in case

$$\left(\sum_{i=1}^2 \lambda_i\right)^2 > 4\pi \sum_{i=1}^2 \xi_i\lambda_i \tag{4.139}$$

(see [29, 146]). The parameter region defined by (4.137) and (4.139) in the quadrant $\lambda_i > 0, i = 1, 2$, of the (λ_1, λ_2)-plane is not empty.

Since property (4.134) means that

$$T < +\infty \implies \lim_{t\uparrow T} \|u(\cdot, t)\|_\infty = +\infty, \tag{4.140}$$

so recalling that $u = \sum_{i=1}^2 u_i$ in (4.131), the blowup set of (u_1, u_2), defined by

$$\mathcal{S} = \{x_0 \in \overline{\Omega} \mid \exists (x_k, t_k) \to (x_0, T), \; u(x_k, t_k) \to +\infty\}, \tag{4.141}$$

is not empty. Similarly to the case of single unknown species (see Sect. 4.5), the formation of collapses occurs, with collapse masses lying on a curve. Henceforth we say that the collapse $m_i(x_0)\delta_{x_0}(dx)$ in (4.142) below is degenerate if $m_i(x_0) = 0$.

Theorem 4.7 ([29, 125]) *If $T < +\infty$, the blowup set $\mathcal{S}$ defined by (4.141) is finite. It holds that*

$$u_i(x, t)dx \to \sum_{x_0 \in \mathcal{S}} m_i(x_0)\delta_{x_0}(dx) + f_i(x)dx, \quad i = 1, 2 \tag{4.142}$$

in $\mathcal{M}(\overline{\Omega}) = C(\overline{\Omega})'$ as $t \uparrow T = T_{\max} < +\infty$, where $m_i(x_0) \geq 0, i = 1, 2$, are constants satisfying $(m_1(x_0), m_2(x_0)) \neq (0, 0)$, and $0 \leq f_i = f_i(x) \in L^1(\Omega), i = 1, 2$, are smooth functions in $\overline{\Omega} \setminus \mathcal{S}$. Moreover,

$$\left(\sum_{i=1}^{2} m_i(x_0)\right)^2 \geq m_*(x_0) \sum_{i=1}^{2} \xi_i m_i(x_0) \tag{4.143}$$

for any $x_0 \in \mathcal{S}$, where

$$m_*(x_0) = \begin{cases} 8\pi, & if\ x_0 \in \Omega \\ 4\pi, & if\ x_0 \in \partial\Omega. \end{cases} \tag{4.144}$$

The following theorem is the first observation to approach collapse mass separation. Here we note that the blowup set $\mathcal{S}$ coincides with the origin for radially symmetric solutions.

Theorem 4.8 ([29, 125]) *Let Ω be a disc centered at the origin, $u_i = u_i(|x|, t)$, $i = 1, 2$, and $T < +\infty$. Then $\mathcal{S} = \{0\}$ and $m_i = m_i(0)$ must satisfy*

$$m_i \leq 8\pi\xi_i, \quad i = 1, 2, \tag{4.145}$$

besides the equality in (4.143) with $x_0 = 0$.

Inequality (4.145) follows from (4.143) if

$$1/2 \leq \xi_i/\xi_j \leq 2, \quad i, j = 1, 2. \tag{4.146}$$

More precisely, if (4.146) holds, then the curve (a parabola if $\xi_1 \neq \xi_2$ and a line in the other case) defined by

$$\left(\sum_{i=1}^{2} m_i\right)^2 = m_* \sum_{i=1}^{2} \xi_i m_i, \quad m_* = m_*(x_0), \tag{4.147}$$

in the quadrant $\{(m_1, m_2) \mid m_i > 0, i = 1, 2\}$ of the (m_1, m_2)-plane does not cross the lines $m_i = \xi_i m_*$, $i = 1, 2$. In the other case, when $\xi_i/\xi_j > 2$ or $\xi_i/\xi_j < 1/2$ for $i \neq j$, one of $(m_1, m_2) = (8\pi\xi_1, 0)$ and $(m_1, m_2) = (0, 8\pi\xi_2)$ is an isolated point of (4.147) in the quadrant $\{(m_1, m_2) \mid m_i \geq 0, i = 1, 2\}$ and (4.145) holds. In view of these observations we can expect that the above isolated endpoint of (m_1, m_2), which we call *mass separation*, may actually exist, and is stable under non-radially symmetric perturbations. In this context we recall that *simultaneous blowup* (4.135) does not always occur for radially symmetric solutions, regardless of the parameter region indicated by (4.137). If both simultaneous blowup and mass separation take place, say, $m_i(x_0) = 0$ in (4.142), then it will hold that $f_i \notin L^\infty(\Omega \cap B(x_0, R))$ for $0 < R \ll 1$, where $B(x_0, R) = \{x \mid |x - x_0| < R\}$.

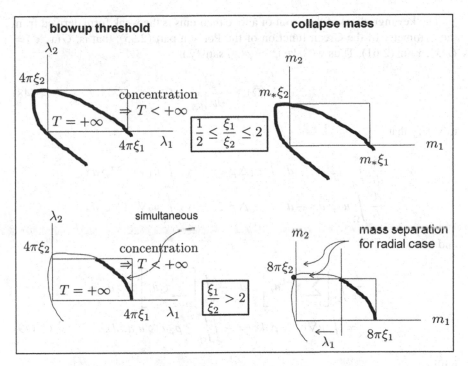

Fig. 4.4 Simultaneous blowup and mass separation

The following theorem shows that the mass separation of radially symmetric solutions actually occurs if the total mass of one component is relatively small compared with that of the other.

Theorem 4.9 ([29, 125]) *Under the assumption of Theorem 4.8, let*

$$\xi_i/\xi_j > 2$$

for some $i \neq j$. Then $m_i = 0$ and hence $m_j = 8\pi\xi_j$ holds, provided that

$$\|u_{i0}\|_1 < 8\pi(\xi_i - 2\xi_j).$$

We note that a sufficient condition for $T < +\infty$ in the above theorem is given in [30], namely,

$$\|u_{j0}\|_1 > 8\pi\xi_j, \quad \||x|^2 u_{j0}\|_1 \ll 1$$

(Fig. 4.4).

The key ingredient of the proof of above theorems is the weak form derived from the symmetry of the Green function of the Poisson part (2.55), that is, $G(x, x') = G(x', x)$ in (2.61). Thus we take $\varphi = \varphi(x)$ satisfying

$$\varphi \in C^2(\overline{\Omega}), \quad \frac{\partial \varphi}{\partial \nu}\bigg|_{\partial\Omega} = 0, \tag{4.148}$$

to verify that

$$\frac{d}{dt} \int_\Omega u_1 \varphi \, dx - d_1 \int_\Omega u_1 \Delta\varphi \, dx = \chi_1 \int_\Omega u_1 \nabla v \cdot \nabla\varphi \, dx,$$

$$\frac{d}{dt} \int_\Omega u_2 \varphi \, dx - d_2 \int_\Omega u_2 \Delta\varphi \, dx = \chi_2 \int_\Omega u_2 \nabla v \cdot \nabla\varphi \, dx,$$

and hence

$$\frac{d}{dt} \int_\Omega \left[\sum_{i=1}^{2} \chi_i^{-1} u_i \right] \varphi \, dx - \int_\Omega \left[\sum_{i=1}^{2} \xi_i u_i \right] \Delta\varphi \, dx$$

$$= \int_\Omega u \nabla v \cdot \nabla\varphi \, dx = \frac{1}{2} \iint_{\Omega\times\Omega} \rho_\varphi u \otimes u \, dx dx', \tag{4.149}$$

with

$$\rho_\varphi(x, x') = \nabla\varphi(x) \cdot \nabla_x G(x, x') + \nabla\varphi(x') \cdot \nabla_{x'} G(x, x').$$

This structure is quite common. For example, the other competitive system of chemotaxis is composed of the Smoluchowski part (4.128)–(4.129) with (4.130), coupled with the Poisson part

$$-\Delta v_1 = u_2 - \frac{1}{|\Omega|} \int_\Omega u_2, \quad \frac{\partial v_1}{\partial \nu}\bigg|_{\partial\Omega} = 0, \quad \int_\Omega v_1 = 0$$

$$-\Delta v_2 = u_1 - \frac{1}{|\Omega|} \int_\Omega u_1, \quad \frac{\partial v_2}{\partial \nu}\bigg|_{\partial\Omega} = 0, \quad \int_\Omega v_2 = 0. \tag{4.150}$$

In this system the first species u_1 secretes a chemical v_1 which attracts the second species u_2, and similarly, u_2 secretes v_2 which attracts u_1. We observe that these hold the fundamental properties of total mass conservation (4.132) and scaling invariance of the limit system

$$\frac{\partial u_i}{\partial t} = d_i \Delta u_i - \chi_i \nabla \cdot u_i \nabla v_i, \quad i = 1, 2,$$

$$-\Delta v_1 = u_2, \quad -\Delta v_2 = u_1 \quad \text{in } \mathbf{R}^2 \times (0, T),$$

that is,

$$u_i^\mu(x, t) = \mu^2 u_i(\mu x, \mu^2 t), \quad i = 1, 2, \quad v_\mu(x, t) = v(\mu x, \mu^2 t)$$

for $\mu > 0$, and the weak form valid for (4.148),

$$\frac{d}{dt} \int_\Omega \left[\sum_{i=1}^2 \chi_i^{-1} u_i \right] \varphi \, dx - \int_\Omega \left[\sum_{i=1}^2 \xi_i u_i \right] \Delta \varphi \, dx$$

$$= \frac{1}{2} \iint_{\Omega \times \Omega} \rho_\varphi u_1 \otimes u_2 \, dx dx',$$

respectively. Fact that the total free energy sis decreasing, i.e.,

$$\frac{d}{dt} \mathcal{F}_{\xi_1, \xi_2}(u_1, u_2) \leq 0,$$

$$\mathcal{F}_{\xi_1, \xi_2}(u_1, u_2) = \sum_{i=1}^2 \int_\Omega \xi_i u_i (\log u_i - 1) dx - \frac{1}{2} \langle (-\Delta)^{-1} u_1, u_2 \rangle,$$

is also to be noted.

We thus obtain the following result.

Theorem 4.10 ([28, 125]) *If $T < +\infty$, the blowup set $\mathcal{S}$ defined by (4.141) is finite. It holds that (4.142) in $\mathcal{M}(\overline{\Omega}) = C(\overline{\Omega})'$ as $t \uparrow T = T_{\max} < +\infty$, where $m_i(x_0) \geq 0$, $i = 1, 2$, are constants satisfying (4.133) and $0 \leq f_i = f_i(x) \in L^1(\Omega)$, $i = 1, 2$, are*

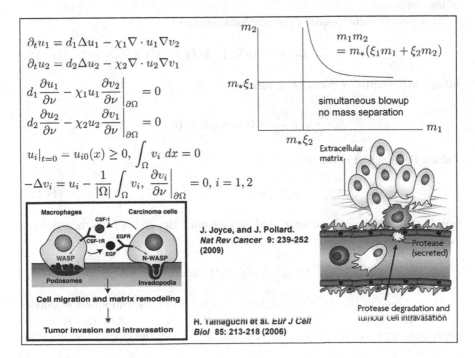

Fig. 4.5 Cross chemotaxis

smooth functions in $\overline{\Omega} \setminus \mathcal{S}$. *We have* $f_i > 0$ *in* $\overline{\Omega} \setminus \mathcal{S}$ *except for* $u_{i0} \equiv 0$. *The collapse masses* $m_i(x_0)$, $i = 1, 2$, *satisfy*

$$m_1(x_0)m_2(x_0) \geq m_*(x_0) \sum_{i=1}^{2} \xi_i m_i(x_0). \tag{4.151}$$

Since (4.151) with $m_i(x_0) \geq 0$, $i = 1, 2$ and $(m_1(x_0), m_2(x_0)) \neq (0, 0)$, it holds that $m_i(x_0) > m_*(x_0)\xi_i$, $i = 1, 2$, by (4.151). Hence, any components of the collapse are non-degenerate. In particular, we have simultaneous blowup without mass separation in this system even for non-radially symmetric solutions (Fig. 4.5).

4.5   Method of the Weak Scaling Limit

We have described the study of chemotaxis systems on bounded domains in two dimensions. Here we show the technical ingredients of the proof, taking the case of a single component with the formation of a chemical potential subject to the Dirichlet boundary condition according to [124]. For such a system the solution is kept bounded near the boundary and hence the blowup set consists of a finite number of interior points.

This system is composed of the Smoluchowski part

$$u_t = \Delta u - \nabla \cdot (u \nabla v) \quad \text{in } \Omega \times (0, T) \tag{4.152}$$

subject to the null-flux boundary condition

$$\frac{\partial u}{\partial \nu} - u \frac{\partial v}{\partial \nu} = 0 \quad \text{on } \partial\Omega \times (0, T), \tag{4.153}$$

and the Poisson part in the form of

$$- \Delta v = u, \quad v|_{\partial\Omega} = 0; \tag{4.154}$$

here $\Omega \subset \mathbf{R}^2$ is a bounded domain with smooth boundary $\partial\Omega$ and ν is the outer unit normal vector. The initial condition is given as

$$u|_{t=0} = u_0(x) > 0 \quad \text{on } \overline{\Omega}, \tag{4.155}$$

where $u_0 = u_0(x)$ is a smooth function. This is one of the simplified models of chemotaxis originated from [20, 56, 66, 89]. A closely related system is the DD (drift-diffusion) model in semi-conductor physics, where the gradient of potentials created by the particle density is repulsive [12], and hence the Poisson part takes the form

$$\Delta v = u, \quad v|_{\partial\Omega} = 0. \tag{4.156}$$

In statistical physics, system (4.152)–(4.155) describes the mean field motion of many self-gravitating Brownian particles [120]. Several studies have been devoted to the case where the domain Ω is the whole space $\mathbf{R}^2$ (see [13]). The case where the Poisson part is subject to the Neumann boundary condition, for example,

$$-\Delta v = u - \frac{1}{|\Omega|}\int_\Omega u\, dx, \quad \frac{\partial v}{\partial \nu}\bigg|_{\partial\Omega} = 0, \quad \int_\Omega v\, dx = 0 \tag{4.157}$$

is also studied in detail (see Sect. 4.4). These arguments are valid even for (4.152)–(4.155) for interior blowup points.

Although boundary blowup points actually arise in the case of (4.157), their blowup mechanism is essentially described by that of interior blowup points. In fact, first, the boundary conditions of (4.153)–(4.154) are reduced to

$$\frac{\partial u}{\partial \nu} = \frac{\partial v}{\partial \nu} = 0 \quad \text{on } \partial\Omega \times (0, T).$$

Then the boundary blowup mechanism of system (4.152)–(4.153) and (4.157) is reduced to the case that Ω is a compact Riemannian surface without boundary, through the reflection using conformal mapping [122].

Concerning (4.152)–(4.155), the monotonicity formula described below is valid by the positivity of v, which guarantees blowup analysis for the boundary blowup point. Then we can exclude the possibility of boundary blowup.

Theorem 4.11 ([124]) *The blowup solution $u = u(x, t)$ in finite time to (4.152)–(4.155) does not take any boundary blowup points. Hence it holds that $S \subset \Omega$ where*

$$S = \{x_0 \in \overline{\Omega} \mid \exists x_k \to x_0, \exists t_k \uparrow T \text{ such that } u(x_k, t_k) \to +\infty\} \tag{4.158}$$

denotes the blowup set.

System (4.152)–(4.154) and (4.157) is derived from the asymptotic expansion to a parabolic-parabolic system of chemotaxis [56]. Here adding constants to v just violates the last condition $\int_\Omega v = 0$ which is required for the unique solvability of (4.157). Hence essentially v may be taken to be positive because only ∇v is involved in the Smoluchowski part, and this renormalization has a reality in regarding v as a concentration of chemical substances. The Poisson part (4.156), however, will be more reasonable from this point of view because we have always $v > 0$ from the maximum principle. Henceforth, $C_i, i = 1, 2, \ldots, 19$, denote positive constants.

Here we note several fundamental facts. First, local-in-time unique existence of the solution to (4.152)–(4.155) is standard, given smooth initial value $u_0 = u_0(x)$. Maximum principle guarantees $u(\cdot, t) > 0$ on $\overline{\Omega}$ and then the total mass conservation $\|u(\cdot, t)\|_1 = \lambda$ follows from the Smoluchowski part as

$$\frac{d}{dt} \int_\Omega u = \int_\Omega \nabla \cdot (\nabla u - u \nabla v) \, dx = \int_{\partial\Omega} \left(\frac{\partial u}{\partial \nu} - u \frac{\partial v}{\partial \nu} \right) dS = 0$$

where dS denotes the area element. Using the parabolic regularity the existence time T of this solution $u(\cdot, t)$ is estimated from below by $\|u_0\|_\infty$, and hence

$$\lim_{t \uparrow T} \|u(\cdot, t)\|_\infty = +\infty$$

occurs if $T < +\infty$. Then the blowup set S defined by (4.158) is not empty. The free energy

$$\mathcal{F}(u) = \int_\Omega u(\log u - 1) \, dx - \frac{1}{2} \langle (-\Delta)^{-1} u, u \rangle,$$

where $v = (-\Delta)^{-1} u$ stands for (4.154), is decreasing that is,

$$\frac{d}{dt} \mathcal{F}(u) = -\int_\Omega u |\nabla(\log u - v)|^2 \, dx \le 0.$$

This functional $\mathcal{F} = \mathcal{F}(u)$ is associated with the Trudinger-Moser inequality in dual form,

$$\inf\{\mathcal{F}(u) \mid u \ge 0, \ \|u\|_1 = 8\pi\} > -\infty.$$

Similarly to the case of (4.157), a blow-up criterion is obtained by the weak form associated with the test function (4.148), that is,

$$\frac{d}{dt} \int_\Omega u\varphi \, dx = \int_\Omega (-\nabla u + u \nabla v) \cdot \nabla\varphi \, dx$$

$$= \int_\Omega u \Delta\varphi \, dx + \frac{1}{2} \iint_{\Omega \times \Omega} \rho_\varphi u \otimes u \, dx dx', \qquad (4.159)$$

where $u \otimes u = u(x, t)u(x', t)$,

$$\rho_\varphi(x, x') = \nabla\varphi(x) \cdot \nabla_x G(x, x) + \nabla\varphi(x') \cdot \nabla_{x'} G(x, x'), \qquad (4.160)$$

and $G(x, x')$ is the Green function of the Poisson part (4.156). Then we obtain the monotonicity formula.

Lemma 4.5.1 (monotonicity formula) *It holds that*

$$\left| \frac{d}{dt} \int_\Omega u\varphi \, dx \right| \le C_1(\lambda) \|\nabla\varphi\|_{C^1(\overline{\Omega})} \tag{4.161}$$

for any $\varphi = \varphi(x)$ satisfying (4.148).

Proof It suffices to show that $\rho_\varphi \in L^\infty(\Omega \times \Omega)$, with

$$\|\rho_\varphi\|_{L^\infty(\Omega \times \Omega)} \le C_2 \|\nabla\varphi\|_{C^1(\overline{\Omega})}. \tag{4.162}$$

This property follows from the boundary behavior of the Green function similarly to (4.157) (see [122]).

First, this $G = G(x, x')$ is smooth in $\overline{\Omega} \times \overline{\Omega} \setminus D$, $D = \{(x, x) \mid x \in \Omega\}$. Further, the interior regularity holds, as

$$G(x, x') = \Gamma(x - x') + K(x, x'),$$
$$K \in C^{1+\theta,\theta}(\Omega \times \overline{\Omega}) \cap C^{\theta,1+\theta}(\overline{\Omega} \times \Omega) \tag{4.163}$$

with $0 < \theta < 1$, where

$$\Gamma(x) = \frac{1}{2\pi} \log \frac{1}{|x|}.$$

Given $x_0 \in \partial\Omega$, we now take the conformal diffeomorphism

$$X \cdot \overline{\Omega \cap B(x_0, 2R)} \longrightarrow \overline{\mathbf{R}_+^2}, \quad 0 < R \ll 1, \tag{4.164}$$

where $\mathbf{R}_+^2 = \{(X_1, X_2) \mid X_2 > 0\}$. Then using

$$X_* = \begin{pmatrix} X_1 \\ -X_2 \end{pmatrix} \quad \text{for} \quad X = \begin{pmatrix} X_1 \\ X_2 \end{pmatrix},$$

we obtain

$$G(x, x') = E(x, x') + K(x, x'),$$
$$K \in C^{1+\theta,\theta} \cap C^{\theta,1+\theta}(\overline{\Omega \cap B(x_0, R)} \times \overline{\Omega \cap B(x_0, R)}), \tag{4.165}$$

with

$$E(x, x') = \Gamma(X - X') - \Gamma(X - X'_*) \tag{4.166}$$

under the transformation (4.164).

At this stage we note

$$\nabla\Gamma(x - x') \cdot (\nabla\varphi(x) - \nabla\varphi(x')) \in L^\infty(\Omega \times \Omega)$$

for any $\varphi = \varphi(x) \in C^2(\overline{\Omega})$, while if this φ is supported near $x = x_0 \in \partial\Omega$ and satisfies $\left.\frac{\partial\varphi}{\partial X_2}\right|_{X_2=0} = 0$, then it holds that

$$|\nabla\Gamma(X - X'_*) \cdot (\nabla\varphi(X) - \nabla\varphi(X'))| \le C_3,$$

(see [122] p. 89.) Now these localized properties are unified as (4.162) through a partition of unity subject to a finite covering of $\overline{\Omega}$.   □

Henceforth, given $x_0 \in \overline{\Omega}$ and $0 < R \ll 1$, $\varphi = \varphi_{x_0,R}$ denotes the C^2-function on $\overline{\Omega}$ with the support radius $2R > 0$, equal to 1 on $\overline{\Omega} \cap B(x_0, R)$, satisfying $\frac{\partial\varphi}{\partial\nu} = 0$ on $\partial\Omega$, and

$$|\nabla\varphi| \le C_4 R^{-1}\varphi^{5/6}, \quad |\nabla^2\varphi| \le C_4 R^{-2}\varphi^{2/3}$$

(see [117, 122] for the construction of such a cut-off function).

The ε-regularity stated below is proven by means of several estimates.

Lemma 4.5.2 (ε-regularity) *There exists $\varepsilon_0 > 0$ such that*

$$\limsup_{t\uparrow T} \|u(\cdot, t)\|_{L^1(\Omega\cap B(x_0,R))} < \varepsilon_0, \; 0 < \exists R \ll 1 \implies x_0 \notin S.$$

Proof First, the Gagliardo-Nirenberg inequality implies that

$$\begin{aligned}
\|u\|_3^3 &= \|\chi_{u\le s}u\|_3^3 + \|\chi_{u>s}u\|_3^3 \\
&\le C_5\|\chi_{u>s}u\|_{H^1(\Omega)}^2\|\chi_{u>s}u\|_1 + s^3|\Omega|
\end{aligned} \tag{4.167}$$

for $s > 1$. Then we use Poincaré-Wirtinger's inequality

$$\mu_2\|w - \overline{w}\|_2^2 \le \|\nabla w\|_2^2, \quad \overline{w} = \frac{1}{|\Omega|}\int_\Omega w, \quad w \in H^1(\Omega)$$

to get

$$\begin{aligned}
\|\chi_{u>s}u\|_{H^1(\Omega)}^2 &\le \|u\|_{H^1(\Omega)}^2 = \|\nabla u\|_2^2 + \|u\|_2^2 \\
&\le \|\nabla u\|_2^2 + 2(\|u - \overline{u}\|_2^2 + \|\overline{u}\|_2^2) \\
&\le \left(1 + \frac{2}{\mu_2}\right)\|\nabla u\|_2^2 + \frac{2}{|\Omega|}\|u\|_1^2,
\end{aligned}$$

where $\mu_2 > 0$ denotes the second eigenvalue of $-\Delta$ under the Neumann boundary condition. It thus turns out that

$$\|u\|_3^3 \leq C_6(\|\nabla u\|_2^2 + \|u\|_1^2)\frac{1}{\log s}\|u \log u\|_1 + s^3 |\Omega|, \tag{4.168}$$

by (4.167) (see also Lemma 4.1 of [122]).

Let $\varphi = \varphi(x) \geq 0$ be a test function satisfying (4.148). First, we have

$$\frac{d}{dt}\int_\Omega u(\log u - 1)\varphi \, dx = \int_\Omega u_t \log u \cdot \varphi \, dx$$
$$= \int_\Omega (-\nabla u + u\nabla v) \cdot \nabla(\varphi \log u) \, dx = -I + II$$

with

$$I = \int_\Omega \nabla u \cdot \nabla(\varphi \log u) \, dx = \int_\Omega \left[u^{-1}|\nabla u|^2 \varphi + (\log u)\nabla u \cdot \nabla \varphi\right] dx$$
$$= \int_\Omega \left[u^{-1}|\nabla u|^2 \varphi - u(\log u - 1)\Delta \varphi\right] dx.$$

Since

$$\left.\frac{\partial v}{\partial \nu}\right|_{\partial \Omega} \leq 0, \tag{4.169}$$

we obtain

$$II = \int_\Omega \left[\varphi \nabla v \cdot \nabla u + (u \log u)\nabla v \cdot \nabla \varphi\right] dx$$
$$\leq \int_\Omega \left[-u\nabla \cdot (\varphi \nabla v) + (u \log u)\nabla v \cdot \nabla \varphi\right] dx$$
$$= \int_\Omega \left[u^2 \varphi + u(\log u - 1)\nabla v \cdot \nabla \varphi\right] dx$$

where the L^1-estimate for (4.156),

$$\|v\|_{W^{1,q}(\Omega)} \leq C_7(q), \quad 1 \leq q < 2, \tag{4.170}$$

is applicable (see [14]). Putting $\varphi = \varphi_{x_0, R}$, we obtain

$$\frac{d}{dt}\int_\Omega u(\log u - 1)\varphi \, dx + \frac{1}{4}\int_\Omega u^{-1}|\nabla u|^2 \varphi \, dx \leq 2\int_\Omega u^2 \varphi \, dx + C_8(R)$$

similarly to the case of (4.157) (see [117] or Lemma 11.3 of [122]). Then a form of the Gagliardo-Nirenberg inequality applied to the first term on the right-hand side shows that there exists $\varepsilon_0 > 0$ such that

$$\limsup_{t \uparrow T} \|u(\cdot, t)\|_{L^1(\Omega \cap B(x_0, R))} < \varepsilon_0$$

$$\implies \limsup_{t \uparrow T} \int_{\Omega \cap B(x_0, R/2)} u \log u \, dx < +\infty \qquad (4.171)$$

for any $0 < R \ll 1$ (see [117] or the proof of Lemma 11.4 of [122]).

The second step it to show that

$$\limsup_{t \uparrow T} \int_{\Omega \cap B(x_0, R)} u(\log u - 1) \, dx < +\infty, \; 0 < R \ll 1$$

$$\implies x_0 \notin \mathcal{S}. \qquad (4.172)$$

First, we take $p > 0$ and a test function $\varphi = \varphi(x) \geq 0$ satisfying (4.148) to derive

$$\frac{1}{p+1} \frac{d}{dt} \int_\Omega u^{p+1} \varphi \, dx = \int_\Omega u_t u^p \varphi \, dx = -\int_\Omega (\nabla u - u \nabla v) \cdot \nabla(u^p \varphi) \, dx$$

$$= \int_\Omega [\{-p u^{p-1} |\nabla u|^2 + u(\nabla v \cdot \nabla u^p)\} \varphi - u^p \nabla u \cdot \nabla \varphi + u^{p+1} \nabla v \cdot \nabla \varphi] dx.$$

It follows that

$$\frac{1}{p+1} \frac{d}{dt} \int_\Omega u^{p+1} \varphi \, dx + \int_\Omega \left[\frac{4p}{(p+1)^2} \left| \nabla u^{\frac{p+1}{2}} \right|^2 \varphi + u^p \nabla u \cdot \nabla \varphi \right] dx$$

$$= \int_\Omega [u(\nabla v \cdot \nabla u^p)\varphi + u^{p+1} \nabla v \cdot \nabla \varphi] \, dx. \qquad (4.173)$$

Here the first term on the right-hand side is treated by $\varphi \geq 0$ and (4.169), that is,

$$\int_\Omega u(\nabla v \cdot \nabla u^p)\varphi \, dx = \frac{1}{p+1} \int_\Omega (\nabla u^{p+1} \cdot \nabla v)\varphi \, dx$$

$$\leq -\frac{1}{p+1} \int_\Omega u^{p+1} \nabla \cdot (\varphi \nabla v) \, dx$$

$$= \frac{1}{p+1} \int_\Omega u^{p+2} \varphi \, dx - \frac{1}{p+1} \int_\Omega u^{p+1} \nabla v \cdot \nabla \varphi \, dx.$$

Hence the right-hand side of (4.173) is estimated from above by

$$\int_\Omega \left[\frac{1}{p+1} u^{p+2}\varphi + \frac{p}{p+1} u^{p+1}\nabla v \cdot \nabla\varphi \right] dx$$

$$= \int_\Omega \left[\frac{1}{p+1} u^{p+2}\varphi - \frac{p}{p+1} u^{p+1} v\Delta\varphi + v\nabla u^{p+1} \cdot \nabla\varphi \right] dx,$$

recalling that $\frac{\partial\varphi}{\partial\nu} = 0$ on $\partial\Omega$.

Putting $\varphi = \varphi_{x_0,R}$, we use this estimate for $p = 1$ and $p = 2$ as in (11.15) and (11.17) of [122], respectively. We thus assume that

$$\limsup_{t\uparrow T} \int_{\Omega\cap B(x_0,R))} u(\log u - 1)\, dx < +\infty$$

and then apply (4.168) with u replaced by $u\varphi^{1/3}$. It follows that

$$\|u\|_{L^3(\Omega\cap B(x_0,R/2))} \leq C_9,$$

and hence

$$\|\nabla v\|_{L^\infty(\Omega\cap B(x_0,R/4))} \leq C_{10}, \tag{4.174}$$

thanks to the elliptic estimate to (4.154) and Morrey's embedding theorem.

For $p \geq 3$ we apply the relations

$$\frac{1}{p+1}\frac{d}{dt}\int_\Omega (u\varphi)^{p+1}dx = \int_\Omega u_t u^p \varphi^{p+1}dx$$

$$= -\int_\Omega (\nabla u - u\nabla v) \cdot \nabla(u^p \varphi^{p+1})\, dx$$

to set up an iteration scheme. Eventually we reach (4.172) (see [117] or Lemma 11.2 of [122]), and the proof is complete. $\qquad\square$

Given the blowup solution $u = u(\cdot, t)$, $0 \leq t < T$, the monotonicity formula (Lemma 4.5.1) together with the total mass conservation $u \geq 0$, $\|u\|_1 = \lambda$ implies the convergence

$$u(x, t)dx \; \rightharpoonup \; \mu(dx, T) \quad \text{in } \mathcal{M}(\overline{\Omega}) = C(\overline{\Omega})' \text{ as } t \uparrow T.$$

Then this property combined with ε-regularity (Lemma 4.5.2) guarantees that

$$\mu(dx, T) = \sum_{x_0 \in S} m(x_0)\delta_{x_0}(dx) + f(x)dx, \tag{4.175}$$

with $0 \le f = f(x) \in L^1(\Omega)$, $m(x_0) \ge \varepsilon_0$, and $\sharp S < +\infty$. Now, by the parabolic regularity, this $f = f(x)$ is smooth in $\overline{\Omega} \setminus S$ and the parabolic-elliptic system (4.152)–(4.154) holds in $\overline{\Omega} \times [0, T] \setminus (\overline{\Omega} \setminus S) \times \{T\}$ in the classical sense. Hence if $f \ge 0$ takes the value zero in $\overline{\Omega} \setminus S$, then u must be identically zero by the strong maximum principle, contradicting (4.155). Therefore, $0 < f = f(x)$ in $\overline{\Omega} \setminus S$.

To prove Theorem 4.11 we use the argument of [29, 122, 128], which may be called the method of weak scaling limit, that is, the use of the weak solution to the scaled system to control the blowup mechanism of the pre-scaled system. The notion of the weak solution was first introduced for the pre-scaled system in [119]. To begin with, we restate the definition and properties of this weak solution for the pre-scaled system studied in this paper, that is, (4.152)–(4.155).

First, $0 \le \mu = \mu(dx, t) \in C_*([0, T], \mathcal{M}(\overline{\Omega}))$ is called a weak solution to (4.152)–(4.155) if there is $0 \le \mathcal{N} = \mathcal{N}(\cdot, t) \in L_*^\infty([0, T], \mathcal{X}')$, which we call the *multiplication operator*, provided with the following properties. Here, $\mathcal{X}$ denotes the linear closed hull in $L^\infty(\Omega \times \Omega)$ of

$$\mathcal{X}_0 = \left\{ \rho_\varphi + \psi \mid \varphi \in C^2(\overline{\Omega}), \left. \frac{\partial \varphi}{\partial \nu} \right|_{\partial\Omega} = 0, \ \psi \in C(\overline{\Omega} \times \overline{\Omega}) \right\}$$

for $\rho_\varphi \in L^\infty(\Omega \times \Omega)$ defined by (4.160), using $\varphi = \varphi(x)$ in (4.148) and the Green function $G = G(x, x')$ to (4.157).

1. For $\varphi = \varphi(x)$ satisfying (4.148) the mapping $t \in [0, T] \mapsto \langle \varphi, \mu(dx, t) \rangle$ is absolutely continuous and it holds that

$$\frac{d}{dt} \langle \varphi, \mu(dx, t) \rangle = \langle \Delta\varphi, \mu(dx, t) \rangle + \frac{1}{2} \langle \rho_\varphi, \mathcal{N}(\cdot, t) \rangle_{\mathcal{X}, \mathcal{X}'} \quad \text{for a.e. } t$$

2. We have

$$\mathcal{N}(\cdot, t)|_{C(\overline{\Omega} \times \overline{\Omega})} = \mu(dx, t) \otimes \mu(dx', t) \quad \text{for a.e. } t.$$

Here $\mathcal{N} \ge 0$ for $\mathcal{N} \in \mathcal{X}'$ indicates the property

$$\left| \langle f, \mathcal{N} \rangle_{\mathcal{X}, \mathcal{X}'} \right| \le \langle g, \mathcal{N} \rangle_{\mathcal{X}, \mathcal{X}'},$$

valid for any $f, g \in \mathcal{X}$ satisfying $|f| \le g$ a.e. in $\Omega \times \Omega$.

Although such a solution is not unique for a prescribed initial measure, the above properties are sufficient to the argument developed in [118]. As a consequence we obtain the following theorem.

Theorem 4.12 ([119]) *A weak solution $\mu(dx, t)$ to (4.152)–(4.154) cannot exist even locally-in-time if*

$$\mu(\{x_0\}, 0) > 8\pi \quad and \quad \lim_{R \downarrow 0} \frac{1}{R^2} \langle |x - x_0|^2 \varphi_{x_0, R}, \mu(dx, 0) \rangle = 0$$

for some $x_0 \in \Omega$.

An advantage of weak solutions, on the other hand, is that they can be generated from uniformly bounded sequences.

Theorem 4.13 ([119]) *Let* $0 \leq \mu_k(dx, t) \in C_*([0, T], \mathcal{M}(\overline{\Omega}))$, $k = 1, 2, \ldots$, *be a sequence of weak solutions to (4.152)–(4.154). Assume*

$$\mu_k(\overline{\Omega}, 0) + \sup_{0 \leq t \leq T} \|\mathcal{N}_k(\cdot, t)\|_{\mathcal{X}'} \leq C_{11} \tag{4.176}$$

with a constant $C_{11} > 0$ *independent of k, where*

$$0 \leq \mathcal{N}_k(\cdot, t) \in L_*^\infty([0, T], \mathcal{M}(\overline{\Omega}))$$

stands for the multiplication operator associated with $\mu_k(dx, t)$. *Then we can extract a subsequence, denoted by the same symbol, such that*

$$\mu_k(dx, t) \rightharpoonup \mu(dx, t) \quad in \ C_*([0, T], \mathcal{M}(\overline{\Omega})),$$

where $\mu = \mu(dx, t)$ *is a weak solution to (4.152)–(4.154).*

The classical solution $u = u(x, t) \geq 0$ is regarded as a weak solution by using

$$\mu(dx, t) = u(x, t)dx,$$
$$\langle f, \mathcal{N} \rangle_{\mathcal{X}, \mathcal{X}'} = \iint_{\Omega \times \Omega} f(x, x')u(x, t)u(x', t)dxdx',$$
$$f = f(x, x') \in \mathcal{X} \subset L^\infty(\Omega \times \Omega).$$

(This u may be an L^1-function in the space variables.) Then the condition (4.176) is reduced to

$$\|u_{0,k}\|_1 \leq C_{12},$$

although these solutions $u_k, k = 1, 2, \ldots$, must share a common existence time. Thus from a sequence of classical solutions with a uniform L^1-bound of the initial values and a common existence time, we can extract a subsequence converging to a weak solution.

In the blowup analysis we apply the above results to the backward self-similar transformation of u, that is,

$$z(y, s) = (T - t)u(x, t), \quad y = (x - x_0)/(T - t)^{1/2}, \quad s = -\log(T - t),$$

where $u = u(x, t)$ is the classical solution to (4.152)–(4.155) with the blowup time $T < +\infty$ and $x_0 \in \mathcal{S}$. The precise definition of the weak solution to the limiting system in the following lemma is given in the proof. Henceforth, $z(y, s)$ is extended by zero where it is not defined. We also put $\mathcal{M}(\mathbf{R}^2) = C_0(\mathbf{R}^2)'$ for $C_0(\mathbf{R}^2) = \{f \in C(\mathbf{R}^2 \bigcup \{\infty\}) \mid f(\infty) = 0\}$, where $\mathbf{R}^2 \bigcup \{\infty\}$ stands for the one-point compactification of $\mathbf{R}^2$.

Lemma 4.5.3 *Any sequence* $\{s_k\}$ *such that* $s_k \uparrow +\infty$ *admits a subseqeunce* $\{s'_k\} \subset$ $\{s_k\}$ *such that*

$$z(y, s + s'_k)dy \rightharpoonup \zeta(dy, s) \quad in \ C_*(-\infty, +\infty, \mathcal{M}(\mathbf{R}^2)) \qquad (4.177)$$

with the limiting measure $\zeta(dy, s)$ *supported on* $\overline{L}$, *where* L *is the whole space* $\mathbf{R}^2$ *and a half space denoted by* $\mathbf{R}^2_+$ *if* $x_0 \in \Omega$ *and* $x_0 \in \partial\Omega$, *respectively. It is a weak solution to*

$$\zeta_s = \Delta\zeta - \nabla \cdot \zeta\nabla(G_0 * \zeta + |y|^2/4) \quad in \ L \times (-\infty, +\infty) \qquad (4.178)$$

subject to the boundary condition

$$\frac{\partial\zeta}{\partial\nu} - \zeta\frac{\partial G_0 * \zeta}{\partial\nu} = 0 \quad on \ \partial L \times (-\infty, +\infty) \qquad (4.179)$$

if $x_0 \in \partial\Omega$, *with a uniformly bounded multiplication operator denoted by*

$$\mathcal{K}(\cdot, s) \in L^\infty_*(-\infty, +\infty, \mathcal{M}(\mathbf{R}^2)),$$

where

$$G_0(y, y') = \begin{cases} \Gamma(y - y'), & x_0 \in \Omega \\ \Gamma(y - y') - \Gamma(y - y'_*), & x_0 \in \partial\Omega \end{cases}$$

and y_* *stands for the reflection of* $y \in L = \mathbf{R}^2_+$ *with respect to* ∂L.

Proof First, the above $z = z(y, s)$ satisfies

$$z_s = \Delta z - \nabla \cdot z\nabla(w + |y|^2/4), \quad w(y, s) = \int_{\Omega_s} G_s(y, y')z(y', s)dy'$$

$$in \ \bigcup_{s > -\log T} \Omega_s \times \{s\},$$

$$\frac{\partial z}{\partial\nu} - z\frac{\partial}{\partial\nu}(w + |y|^2/4) = 0 \quad on \ \bigcup_{s > -\log T} \partial\Omega_s \times \{s\},$$

where $\Omega_s = (T - t)^{1/2}(\Omega - \{x_0\})$ and $G_s(y, y') = G(x, x')$. Then we use the weak form of this scaled system. In the case when $x_0 \in \partial\Omega$ we take the conformal diffeomorphism (4.164) $X = X(x)$ and put $Y(y, s) = e^{s/2}X(e^{-s/2}y + x_0)$, $y \in e^{s/2}(\Omega \cap B(x_0, R) - \{x_0\})$. Given

$$\varphi \in C^2_0(\overline{\mathbf{R}^2_+}), \quad \frac{\partial\varphi}{\partial\nu}\bigg|_{\partial\mathbf{R}^2_+} = 0, \qquad (4.180)$$

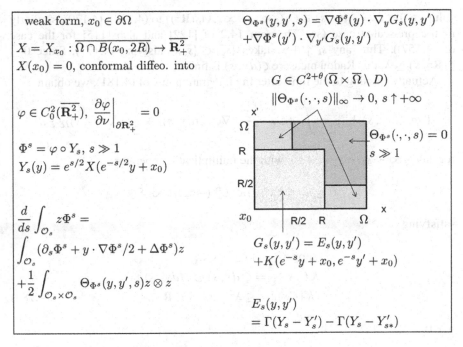

Fig. 4.6 Rescaled Green function

we take $\Phi^s = \varphi \circ Y(\cdot, s)$. This $\Phi^s = \Phi^s(y)$ satisfies

$$\frac{d}{ds}\int_{\mathcal{O}_s} z\Phi^s = \int_{\mathcal{O}_s}(\partial_s \Phi^s + y \cdot \nabla\Phi^s/2 + \Delta\Phi^s)z$$
$$+\frac{1}{2}\iint_{\mathcal{O}_s \times \mathcal{O}_s} \Theta_{\Phi^s}(y, y', s)z \otimes z \qquad (4.181)$$

for $\mathcal{O}_s = \Omega_s \times \{s\}$ and

$$\Theta_{\Phi^s}(y, y', s) = \nabla\Phi^s(y) \cdot \nabla_y G_s(y, y') + \nabla\Phi^s(y') \cdot \nabla_{y'} G_s(y, y')$$

(see (14.14) of [122]).

The asymptotics of $\partial_s \Phi^s + y \cdot \nabla\Phi^s/2 + \Delta\Phi^s$ and $\Theta_{\Phi^s}(y, y', s)$ as $s \uparrow +\infty$ were examined in detail (see the proof of Theorem 14.2 of [122] and also [115]). We have, furthermore,

$$\|z(\cdot, s)\|_{L^1(\Omega_s)} = \lambda, \quad \|z(\cdot, s) \otimes z(\cdot, s)\|_{L^1(\Omega_s \times \Omega_s)} = \lambda^2$$

and the behavior of the Green function (4.165) for $x_0 \in \partial\Omega$ (Fig. 4.6), and (4.163) for $x_0 \in \Omega$. These properties are sufficient to guarantee the convergence to a weak solution $0 \leq \zeta = \zeta(dy, s) \in C_*(-\infty, +\infty, \mathcal{M}_0(\mathbf{R}^2))$ to (4.178)–(4.179), similarly to the pre-scaled case (see Theorem 14.2 of [122] and also [115] for the case of (4.157)). Thus any $s_k \uparrow +\infty$ takes $\{s'_k\} \subset \{s_k\}$ satisfying (4.177). Since $0 \leq \zeta(\mathbf{R}^2, s) \leq \lambda$, this Radon measure $\zeta(dy, s)$ is finite.

Actually, passing to the limit after the integration in s of (4.181), we obtain

$$[\langle \varphi, \zeta(dy, s) \rangle]_{s=s_1}^{s=s_2} = \int_{s_1}^{s_2} \langle \Delta \varphi + \frac{y}{2} \cdot \nabla \varphi, \zeta(dy, s) \rangle + \frac{1}{2} \langle \rho_\varphi^0, \mathcal{K}(\cdot, s) \rangle_{\mathcal{E}, \mathcal{E}'} ds$$

(4.182)

for any $-\infty < s_1 < s_2 < +\infty$, with the multiplication operator

$$0 \leq \mathcal{K} = \mathcal{K}(\cdot, s) \in L^\infty_*(-\infty, +\infty; \mathcal{E}')$$

satisfying

$$\text{supp } \mathcal{K}(\cdot, s) \subset \overline{L} \times \overline{L},$$
$$\mathcal{K}(\cdot, s)|_y = \zeta(dy, s) \otimes \zeta(dy', s),$$
$$\|\mathcal{K}(\cdot, s)\|_{\mathcal{E}'} \leq \lambda^2 \quad \text{a.e. } s \in \mathbf{R},$$

(4.183)

where

$$\rho_\varphi^0(y, y') = \nabla \varphi(y) \cdot \nabla_y G_0(y, y') + \nabla \varphi(y') \cdot \nabla_{y'} G_0(y, y') \in L^\infty(L \times L).$$

Here we have used the following function spaces:

$$C_0^2(\overline{L}) = \{\varphi \in C^2(\overline{L}) \mid \text{supp } \varphi : \text{compact }, \left. \frac{\partial \varphi}{\partial \nu} \right|_{\partial L} = 0\},$$
$$\mathcal{E} = \text{closed linear hull of } \mathcal{E}_0 \text{ in } L^\infty(L \times L),$$
$$\mathcal{E}_0 = \{\rho_\varphi^0 + \psi \mid \varphi \in C_0^2(\overline{L}), \; \psi \in \mathcal{Y}\},$$
$$\mathcal{Y} = C_0(\overline{L} \times \overline{L}) \oplus [(C_0(\overline{L}) \oplus \mathbf{R}) \otimes \mathbf{R}] \oplus [\mathbf{R} \otimes (C_0(\overline{L}) \oplus \mathbf{R})],$$

with $C_0(\overline{L} \times \overline{L})$ standing for the set of continuous functions on $(\overline{L} \cup \{\infty\}) \times (\overline{L} \cup \{\infty\})$ vanishing on $\overline{L} \times \{\infty\} \bigcup \{\infty\} \times \overline{L}$, and $C_0(\overline{L})$ for the set of continuous functions on $\overline{L} \cup \{\infty\})$ vanishing at ∞.

From (4.182) it follows that the mapping $s \in (-\infty, +\infty) \mapsto \langle \varphi, \zeta(dy, s) \rangle$ is locally absolutely continuous for any φ in (4.180), and

$$\frac{d}{ds} \langle \varphi, \zeta(dy, s) \rangle = \langle \Delta \varphi + \frac{y}{2} \cdot \nabla \varphi, \zeta(dy, s) \rangle + \frac{1}{2} \langle \rho_\varphi^0, \mathcal{K}(\cdot, s) \rangle_{\mathcal{E}, \mathcal{E}'}$$

(4.184)

for a.e. $s \in \mathbf{R}$. These properties mean that $\zeta(dy, s)$ is a weak solution to (4.178). $\square$

The parabolic envelope now works similarly to [29, 122, 128] for problem (4.157). It is the key ingredient of the argument.

Lemma 4.5.4 (parabolic envelope) *It holds that*

$$\zeta(\mathbf{R}^2, s) = m(x_0) \tag{4.185}$$

and

$$I(s) \equiv \langle |y|^2, \zeta(dy, s) \rangle \le C_{13} \tag{4.186}$$

for all $s \in (-\infty, +\infty)$.

Proof First, we take $\varphi = \varphi_{x_0, R}$, $0 < R \le 1$, in Lemma 4.5.1. We have

$$|\langle \varphi_{x_0, R}, \mu(dx, T) \rangle - \langle \varphi_{x_0, R}, \mu(dx, t) \rangle| \le \int_t^T \left| \frac{d}{dt} \int_\Omega u \varphi_{x_0, R} dx \right| dt$$

$$\le (T - t) C_1(\lambda) \|\nabla \varphi_{x_0, R}\|_{C^1(\overline{\Omega})} \le C_{14}(T - t)/R^2, \quad 0 \le t < T.$$

Since this estimate holds for any $0 < R \le 1$ and $0 \le t < T$, we can show $R = b(T - t)$ for $0 < T - t \ll 1$, given $b > 0$:

$$|\langle \varphi_{x_0, b(T-t)}, \mu(dx, T) \rangle - \langle \varphi_{x_0, b(T-t)}, \mu(dx, t) \rangle| \le C_{14}/b^2, \quad 0 < T - t \ll 1$$

which means that

$$|\langle \varphi_{x_0, be^{-s}}, \mu(dx, T) \rangle - \langle \varphi_{0, b}, z(y, s) dy \rangle| \le C_{14}/b^2, \quad s \gg 1.$$

Here we shift s to $s + s'_k$ for fixed $s \in (-\infty, +\infty)$ to make $k \to \infty$ and then $b \uparrow +\infty$. It follows that (4.185) from (4.175).

The proof of (4.186) is similar. If $x_0 \in \Omega$ is this case we take $\varphi = |x - x_0|^2 \varphi_{x_0, R}$, $0 < R \ll 1$. Then it follows that

$$\left| \frac{d}{dt} \int_\Omega |x - x_0|^2 \varphi_{x_0, R} u \, dx \right| \le C_{15}$$

by Lemma 4.5.1, which implies the desired estimate (4.186) similarly. If $x_0 \in \partial\Omega$, we just take $\varphi = |X|^2 \varphi_{x_0, R}$, using the conformal change of variable $X = X(x)$ introduced in the previous section (see also [115]). □

We are ready to give the

Proof of Theorem 4.11 Given $x_0 \in \partial\Omega \cap S$, we obtain $\zeta = \zeta(dy, s)$ satisfying (4.184), (4.185), and (4.186) for $G_0 = G_0(y, y')$ defined by

$$G_0(y, y') = \Gamma(y - y') - \Gamma(y - y'_*).$$

Here we take a radially symmetric C^2-function $\psi = \psi(y)$ with support of radius 2, equal to 1 in $|y| \leq 1$, and $0 \leq \psi \leq 1$ and put $\varphi = |y|^2 \psi_R$, $\psi_R(y) = \psi(y/R)$. This φ satisfies (4.180), and hence (4.184) or (4.182) is available:

$$\int_{s_1}^{s_2} \left[\langle \Delta\varphi + \frac{y}{2} \cdot \nabla\varphi, \zeta(dy, s) \rangle + \frac{1}{2} \langle \rho_\varphi^0, \mathcal{K}(\cdot, s) \rangle_{\mathcal{E}, \mathcal{E}'} \right] ds = \left[\langle \varphi, \zeta(dy, s) \rangle \right]_{s=s_1}^{s=s_2}$$

where $-\infty < s_1 < s_2 < +\infty$.

Taking the limit as $R \uparrow +\infty$ in this equality, first, we use the dominated convergence theorem for the first term on the left-hand side. This yields

$$\lim_{R \uparrow +\infty} \int_{s_1}^{s_2} \langle \Delta\varphi + \frac{y}{2} \cdot \nabla\varphi, \zeta(dy, s) \rangle ds = 4(s_2 - s_1)m(x_0) + \int_{s_1}^{s_2} I(s)ds,$$

since

$$\Delta|y|^2 = 4, \quad y \cdot \nabla|y|^2 = 2|y|^2 \tag{4.187}$$

and (4.185)–(4.186).

To control the second term on the left-hand side of (4.182), we use that $\mathcal{K}(\cdot, s) \geq 0$ and $\mathcal{K}(\cdot, s)|_y = \zeta(dy, s) \otimes \zeta(dy', s)$ for a.e. s. First, we have

$$\rho_{|y|^2}^0(y, y') = 0, \tag{4.188}$$

because

$$\nabla\Gamma(y - y') \cdot (\nabla|y|^2 - \nabla|y'|^2) = \frac{1}{\pi},$$

$$\nabla\Gamma(y - y_*') \cdot \nabla|y|^2 + \nabla\Gamma(y' - y_*) \cdot \nabla|y'|^2 = \frac{1}{\pi},$$

recalling that

$$(y - y_*') \cdot y + (y' - y_*) \cdot y' = |y|^2 - 2y \cdot y_*' + |y_*|^2 = |y - y_*'|^2 = |y' - y_*|^2.$$

For $\varphi = |y|^2 \psi_R$, next, it holds that

$$\begin{aligned}
\rho_\varphi^0(y, y') &= \nabla(|y|^2 \psi_R(y)) \cdot \nabla_y G_0(y, y') + \nabla(|y'|^2 \psi_R(y')) \cdot \nabla_{y'} G_0(y, y') \\
&= \left\{ |y|^2 \nabla\psi_R(y) \cdot \nabla_y G_0(y, y') + |y'|^2 \nabla\psi_R(y') \cdot \nabla_{y'} G_0(y, y') \right\} \\
&\quad + \left\{ \psi_R(y)\nabla|y|^2 \cdot \nabla_y G_0(y, y') + \psi_R(y')\nabla|y'|^2 \cdot \nabla_{y'} G_0(y, y') \right\} \\
&= I + II.
\end{aligned} \tag{4.189}$$

We shall treat the second term on the right-hand side of (4.189) by distinguishing three parts of $L \times L$ in the (y, y')-plane. First, for $|y| < 2R$, $y \in \overline{L}$ we use

$$II = (\psi_R(y) - \psi_R(y'))\nabla|y|^2 \cdot \nabla_y G_0(y, y'), \tag{4.190}$$

as derived from (4.188). It then follows that

$$|II| \le \|\nabla\psi\|_\infty \cdot \frac{|y - y'|}{R} \cdot 2|y| \cdot \frac{1}{\pi|y - y'|} \le \frac{2}{\pi}\|\nabla\psi\|_\infty \varphi_{0,4R}(y) \cdot \frac{|y|}{R}, \quad y, y' \in \overline{L},$$

since

$$\frac{|y - y'|}{|y - y'_*|} \le 1, \quad y, y' \in \overline{L}.$$

A similar estimate holds for $|y'| < 2R$ and hence

$$|II| \le C_{16}(\varphi_{0,4R}(y)\frac{|y|}{R} + \varphi_{0,4R}(y')\frac{|y'|}{R}), \quad y, y' \in \overline{L},$$

because $II = 0$ is obvious for $|y| \ge 2R$ and $|y'| \ge 2R$.

Now the property $\mathcal{K}(\cdot, s) \ge 0$ ensures that

$$|\langle II, \mathcal{K}(\cdot, s)\rangle_{\varepsilon,\varepsilon'}| \le \langle C_{16}(\varphi_{0,4R}(y)\frac{|y|}{R} + \varphi_{0,4R}(y')\frac{|y'|}{R}), \mathcal{K}(\cdot, s)\rangle_{\varepsilon,\varepsilon'}$$

$$= C_{16}\langle(\varphi_{0,4R}(y)\frac{|y|}{R} + \varphi_{0,4R}(y')\frac{|y'|}{R}, \zeta(dy, s) \otimes \zeta(dy', s)\rangle \tag{4.191}$$

together with $\mathcal{K}(\cdot, s)|_y = \zeta(dy, s) \otimes \zeta(dy', s)$. The right-hand side on (4.191) is bounded in both R and s, while for each s it converges to 0 as $R \uparrow +\infty$ by the dominated convergence theorem for the product measure $\zeta(dy, s) \otimes \zeta(dy', s)$. Hence it holds that

$$\lim_{R\uparrow+\infty} \int_{s_1}^{s_2} \langle II, \mathcal{K}(\cdot, s)\rangle_{\varepsilon,\varepsilon'} ds = 0, \tag{4.192}$$

again by the dominated convergence theorem.

We use the four parts of $L \times L$ for the first term on the right-hand side of (4.189) to treat. First, we divide I as

$$I = (|y|^2 - |y'|^2)\nabla\psi_R(y) \cdot \nabla_y G_0(y, y')$$
$$+ |y'|^2(\nabla\psi_R(y) - \nabla\psi_R(y')) \cdot \nabla_{y'} G_0(y, y')$$
$$= III + IV$$

for $y, y' \in \overline{L}$. Gir $|y| < 4R$ and $|y'| < 4R$, we have

$$|III| \leq C_{17} \left(\varphi_{0,8R}(y)|y| + \varphi_{0,8R}(y')|y'| \right) \cdot \frac{|y|}{R},$$

$$|IV| \leq C_{17} \left(\varphi_{0,8R}(y') + \varphi_{0,8R}(y) \right) \cdot \frac{|y'|^2}{R^2},$$

while it is obvious that $III = IV = 0$ for $|y| \geq 2R$ and $|y'| \geq 2R$.
   If $|y| < 2R$ and $|y'| \geq 4R$, on the other hand, we come back to

$$I = |y|^2 \nabla \psi_R(y) \cdot \nabla_y G_0(y, y') + |y'|^2 \nabla \psi_R(y') \cdot \nabla_{y'} G_0(y, y').$$

Since $|y - y'| \geq 2R$ holds in this case, we have

$$\left| |y|^2 \nabla \psi_R(y) \cdot \nabla_y G_0(y, y') \right| \leq C_{18} \varphi_{0,4R}(y) \frac{|y|^2}{R^2},$$

$$|y'|^2 \nabla \psi_R(y') \cdot \nabla_{y'} G_0(y, y') = 0.$$

Similarly, for $|y| \geq 4R$ and $|y'| < 2R$ we have

$$|y|^2 \nabla \psi_R(y) \cdot \nabla_y G_0(y, y') = 0,$$

$$\left| |y'|^2 \nabla \psi_R(y') \cdot \nabla_{y'} G_0(y, y') \right| \leq C_{19} \varphi_{0,4R}(y') \frac{|y'|^2}{R^2}.$$

Thus we end up with

$$|I| \leq C_{20} \left(\varphi_{0,8R}(y)(1 + |y|) + \varphi_{0,8R}(y')(1 + |y'|) \right) \times$$

$$\times \left(\frac{|y|}{R} + \frac{|y|^2}{R^2} + \frac{|y'|}{R} + \frac{|y'|^2}{R^2} \right), \quad y, y' \in \overline{L}$$

which implies that

$$\lim_{R \uparrow +\infty} \int_{s_1}^{s_2} \langle I, \mathcal{K}(\cdot, s) \rangle_{\mathcal{E}, \mathcal{E}'} ds = 0, \qquad (4.193)$$

again by the dominated convergence theorem, recalling that

$$\langle 1 + |y|^2, \zeta(dy, s) \rangle \leq C_{13}$$

obtained by (4.185)–(4.186) (Fig. 4.7).
   These properties are summarized as

$$\lim_{R \uparrow +\infty} \int_{s_1}^{s_2} \langle \rho_\varphi^0, \mathcal{K}(\cdot, s) \rangle_{\mathcal{E}, \mathcal{E}'} ds = 0,$$

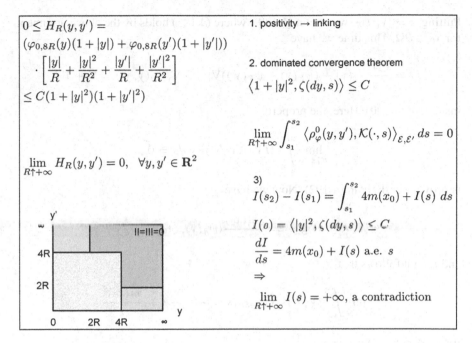

Fig. 4.7 Weak scaling limit

and hence

$$I(s_2) - I(s_1) = \int_{s_1}^{s_2} 4m(x_0) + I(s)\,ds, \quad -\infty < s_1 < s_2 < \infty.$$

The mapping $s \in (-\infty, +\infty) \mapsto I(s) = \langle |y|^2, \zeta(dy, s) \rangle$ is thus locally absolutely continuous and it holds that

$$\frac{dI}{ds} = 4m(x_0) + I \quad \text{for a.c. } s. \tag{4.194}$$

However, since $m(x_0) \geq \varepsilon_0$, (4.194) implies

$$\lim_{s \uparrow +\infty} I(s) = +\infty.$$

This contradicts (4.186) and so we conclude that $\partial \Omega \cap S = \emptyset.$ $\qquad \square$

Given $x_0 \in \Omega$, we have $G_0(y, y') = \Gamma(y - y')$ and hence

$$\rho^0_{|y|^2}(y, y') = -\frac{1}{2\pi}.$$

Putting $\varphi = |y|^2 \psi_R$, we obtain (4.189), where (4.193) holds by the above argument for $x_0 \in \partial\Omega$. This time we have

$$II = -\frac{1}{\pi}\psi_R(y) + (\psi_R(y) - \psi_R(y'))\nabla|y|^2 \cdot \nabla_y G_0(y, y') = V + VI$$

instead of (4.190). Here, the property

$$\lim_{R\uparrow+\infty} \int_{s_1}^{s_2} \langle IV, \mathcal{K}(\cdot, s)\rangle_{\mathcal{E},\mathcal{E}'} ds = 0$$

is proven similarly to (4.192). Now we have

$$\langle V, \mathcal{K}(\cdot, s)\rangle_{\mathcal{E},\mathcal{E}'} = -\frac{m(x_0)}{\pi}\langle \psi_R(y), \zeta(dy, s)\rangle_{\mathcal{E},\mathcal{E}'},$$

and then it follows that

$$\lim_{R\uparrow+\infty} \int_{s_1}^{s_2} \langle V, \mathcal{K}(\cdot, s)\rangle_{\mathcal{E},\mathcal{E}'} ds = -(s_2 - s_1) \cdot \frac{m(x_0)^2}{\pi}$$

for $-\infty < s_1 < s_2 < +\infty$.

These relations guarantee that $s \in (-\infty, +\infty) \mapsto I(s)$ is again locally absolutely continuous and also

$$\frac{dI}{ds} = 4m(x_0) - \frac{m(x_0)^2}{2\pi} + I \quad \text{for a.e. } s \tag{4.195}$$

instead of (4.194). Then by (4.186) we have

$$4m(x_0) - \frac{m(x_0)^2}{2\pi} \leq 0, \tag{4.196}$$

which implies $m(x_0) \geq 8\pi$ from $m(x_0) \geq \varepsilon_0$.

To describe the detailed blowup mechanism we use the scaling back

$$A(dy', s') = e^s \zeta(dy, s), \quad y' = e^{-s/2}y, \quad s' = -e^{-s}$$

to obtain a weak solution $A(dy', s')$ to

$$A_s = \Delta A - \nabla \cdot A\nabla\Gamma * A \quad \text{in } \mathbf{R}^2 \times (-\infty, 0), \tag{4.197}$$

which is formulated similarly. Here we emphasize that

$$0 \leq A = A(dy, s) \in C_*(-\infty, 0, \mathcal{M}(\mathbf{R}^2))$$

is again a finite measure associated with a uniformly bounded multiplication operator, and satisfies

$$A(dy, s) = m(x_0), \quad s \leq 0.$$

The fundamental tool to study (4.197) is the following.

Theorem 4.14 (Liouville property) *Any weak solution* $B = B(dx, t)$ *to*

$$B_t = \Delta B - \nabla \cdot B \nabla \Gamma * B \quad in \, \mathbf{R}^2 \times (-\infty, +\infty) \tag{4.198}$$

satisfies either $B(\mathbf{R}^2, t) = 8\pi, \forall t$, *or* $B(\mathbf{R}^2, t) = 0, \forall t$.

If $I(t) = \langle |x|^2, B(dx, t) \rangle$ converges, we can show that

$$\frac{dI}{dt} = 4M - \frac{M^2}{2\pi}, \quad -\infty < t < +\infty$$

for $M = B(\mathbf{R}^2, t)$. Then we have $M \equiv 0$ or $M \equiv 8\pi$ since $I(t) \geq 0, \forall t$. For the general case we apply the argument of [71] using the local second moment and scaling invariance.

We apply Theorem 4.14 for the translation limit of $A(dy, s)$ in (4.197). Thus taking $s_k \uparrow +\infty$, we apply the *concentration compactness principle* of [82] to $\{A(dy, -s_k)\}$. Hence if this sequence is compact then we have $m(x_0) = 8\pi$. When it is vanishing, there arises a type-I blowup rate. Continuing this process, a list of blowup mechanisms is obtained [125].

From the last term $|y|^2/4$ of (4.178), the regular part of $\zeta(dy, s)$ is attracted to $y = \infty$ as $s \uparrow +\infty$, while (4.186) indicates that the mass of $\zeta(dy, s)$ is mostly enclosed in the prescribed finite region of the variable $y \in \mathbf{R}^2$. This discrepancy implies the residual vanishing, which means that $\zeta(dy, s + s_k)$ decomposes into a sum of quantized delta functions as $s_k \uparrow +\infty$ ([127]), which implies $m(x_0) = 8\pi m$, $m \in \mathbf{N}$, in (4.175) (Theorem 1.2.2 of [123]). Here, the case $m \geq 2$ indicates the collision of subcollapses. The actual possibility of this phenomenon has been suspected based on formal asymptotic analysis [83]. Rigorously, on the other hand, it is proven that $m \geq 2$ occurs if and only if the free energy is unbounded, that is, $\lim_{t \uparrow T} \mathcal{F}(u(\cdot, t)) = -\infty$ (Theorem 1.10.2 of [123]).

Bibliography

1. Adams, E.E., L.W. Gelhar.: Field study of dispersion in a heterogeneous aquifer
2. Allen, S.M., Cahn, J.W.: A macroscopic theory for antiphase boundary motion and its application to antiphase domain coarsening. Acta Metall. **27**, 1085–1095 (1979)
3. Alikakos, N.D.: An application of the invariance principle to reaction-diffusion equations. J. Differ. Equ. **33**, 201–225 (1979)
4. Anderson, A.R.A., Chaplain, M.A.J.: Continuous and discrete mathematical models of tumor-induced angiogenesis. Bull. Math. Biol. **60**, 857–899 (1998)
5. Anderson, A.R.A., Pitcairn, A.W.: Application of the hybrid discrete-continuous technique. In: Alt, W., Chaplain, M., Grebel, M., Lenz, J. (eds.) Polymer and Cell Dynamics-Multiscale Modeling and Numerical Simulations, pp. 261–279. Birkhäuser, Basel (2003)
6. Andrews, S.S., Bray, D.: Stochastic simulation of chemical reactions with spatial resolution and single molecule detail. Phys. Biol. **1**, 137–151 (2004)
7. Berkowitz, B., Cortis, A., Dentz, M., Scher, H.: Modeling non-Fickian transport in geological formations as continous time random walk. Rev. Geophys. **44**(RG2003), 1–49 (2006)
8. Berkowitz, B., Scher, H.: Anomalous transporting laboratory–scale, heterogeneous porous media. Water Resour. Res. **36**, 149–158 (2000)
9. Berry, H.: Monte Carlo simulations of enzyme reactions in two dimensions: fractal kinetics and spatial segregation. Biophys. J. **83**, 1891–1901 (2002)
10. Biler, P.: Local and global solvability of some parabolic systems modelling chemotaxis. Adv. Math. Sci. Appl. **8**, 715–743 (1998)
11. Biler, P., Corrias, P., Dolbeault, J.: Large mass self-similar solutions of the parabolic-parabolic Keller-Segel model of chemotaxis. J. Math. Biol. **63**, 1–32 (2011)
12. Biler, P., Hebisch, W., Nadzieja, T.: The Debye system: existence and large time behavior of solutions. Nonlinear Anal. **23**, 1189–1209 (1994)
13. Blanchet, A., Dolbeaut, J., Perthame, B.: Two-dimensional Keller-Segel model: optimal critical mass and qualitative properties of the solutions. Electron. J. Differ. Equ. **2006-44**, 1–33 (2006)
14. Brezis, H., Strauss, W.: Semi-linear second-order elliptic equations in L^1. J. Math. Soc. Japan **25**, 565–590 (1973)
15. Caginalp, G.: An analysis of a phase field model of a free boundary. Arch. Rational Mech. Anal. **92**, 990–1008 (1993)
16. Cahn, J.W., Hilliard, J.E.: Free energy of a nonuniform system I. Interfacial free energy. J. Chem. Phys. **28**, 258–267 (1958)
17. Cannon, J.R., Hill, C.D.: On the movement of a chemical reaction interface. Indiana Univ. Math. J. **20**, 429–454 (1970)

18. Chaplain, M.J.A., Anderson, A.R.A.: Mathematical modelling of tissue invasion. In: Preziosi, L. (ed.) Cancer Modelling and Simulation, pp. 269–297. Chapman and Hall/CRC, Boca Raton (2003)

19. Chavanis, P.H.: Generalized kinetic equations and effective thermodynamics. Banach Center Publ. **66**, 79–101 (2004)

20. Childress, S., Percus, J.K.: Nonlinear aspects of chemotaxis. Math. Biosci. **56**, 217–237 (1981)

21. Conca, C., Espejo, E.E., Vilches, K.: Remarks on the blow-up and global existence for a two-species chemotactic Keller-Segel system in R^2. Eur. J. Appl. Math. **22**, 553–580 (2011)

22. Corrias, L., Perthame, B., Zaag, H.: A chemotaxis model motivated by angiogenesis. C.R. Acad. Sci. Paris Ser. I **336**, 141–146 (2003)

23. Corrias, L., Perthame, B., Zaag, H.: Global solutions of some chemotaxis and angiogenesis systems in high space dimensions. Milan J. Math. **72**, 1–28 (2004)

24. Dancer, E.N.: On stability and Hopf bifurcations for chemotaxis systems. Math. Appl. Anal. **8**, 245–256 (2001)

25. De Mottoni, P., Rothe, F.: Convergence to homogeneous equilibrium state for generalized Volterra-Lotka systems with diffusion. SIAM J. Appl. Math. **37**, 648–663 (1979)

26. den Hartigh, J.C., van, : Bergen en Henegouwen, P.M., Verkleij, A.J. Boonstra, J. The EGF receptor is an actin-binding protein. J. Cell Biol. **119**, 349–355 (1992)

27. Erban, R., Chapman, S.J.: Stochastic modeling of reaction-diffusion processes: algorithms for bimolecular reactions. Phys. Biol. **6**, 046001 (2009)

28. Espejo, E.E., Kurokiba, M., Suzuki, T.: Blowup threshold and collapse mass separation for a drift-diffusion system in space-dimension two. Comm. Pure Appl. Anal. **12**, 2627–2644 (2013)

29. Espejo, E.E., Stevens, A., Suzuki, T.: Simultaneous blowup and mass separation during collapse in an interacting system of chemotactic species. Diff. Integr. Equ. **25**, 251–288 (2012)

30. Espejo, E.E., Stevens, A., Velázquez, J.J.L.: Simultaneous finite time blow-up in a two-species model for chemotaxis. Analysis **29**, 317–338 (2009)

31. Espejo, E.E., Stevens, A., Velázquez, J.J.L.: A note on non-simultaneous blow-up for a drift-diffusion model. Diff. Integr. Equ. **23**, 451–462 (2010)

32. Evans, L.C.: A convergence theorem for a chemical diffusion-reaction system. Houston J. Math. **6**, 259–267 (1980)

33. FitzHugh, R.: Impulses and physiological states in theoretical models of nerve membrane. Biophys. J. **1**, 445–466 (1961)

34. Fix, G.J.: Phase field method for free boundary problems. In: Fasano, A., Primicerio, M. (eds.) Free Boundary Problems, pp. 580–589. Pitmann, London (1983)

35. Friedman, A., Tello, J.I.: Stability of solutions of chemotaxis equations in reinforced random walks. J. Math. Anal. Appl. **272**, 138–163 (2002)

36. Gajewski, H., Zacharias, K.: Global behaviour of a reaction-diffusion system modelling chemotaxis. Math. Nachr. **195**, 77–114 (1998)

37. Gallinato, O., Ohta, M., Poignard, C., Suzuki, T.: Free boundary problem for cell protrusion formations: theoretical and numerical aspects, J. Math. Biol. (in press)

38. Gierer, A., Meinhardt, H.: A theory of biological pattern formation. Kybernetik **12**, 30–39 (1972)

39. Hale, J.K.: Asymptotic Behavior of Dissipative Systems. American Mathematical Soc, Providence, RI (1988)

40. Halperin, B.I., Hohenberg, P.C., Ma, S.K.: Renormalization-group methods for critical dynamics: I, Recursion relations and effects of energy conservation. Phys. Rev. B **10**, 139–153 (1974)

41. Hatano, Y., Hatano, N.: Dispersive transport of ions in column experiments: an explanation of long-tailed profiles. Water Resour. Res. **34**, 1027–1033 (1998)

42. Henry, D.: Geometric Theory of Semilinear Parabolic Equations. Lecture Notes in Math, vol. 840. Springer, Heidelberg (1981)

43. Herrero, M.A., Velázquez, J.J.L.: Singularity patterns in a chemotaxis model. Math. Ann. **306**, 583–623 (1996)

44. Hilfer, R., Anton, L.: Fractional master equation and fractal time random walk. Phys. Rev. E **51**, 848–851 (1995)
45. Hohenberg, P.C., Halperin, B.I.: Theory of dynamic critical phenomena. Rev. Mod. Phys. **49**, 435–479 (1977)
46. Horstmann, D.: Generalizing Keller-Segel: Lyapunov functionals, steady state analysis and blow-up results for multi-species chemotaxis in the presence of attraction and repulsion between competitive interacting species. J. Nonlinear Sci. **21**, 231–270 (2011)
47. Horstmann, D., Lucia, M.: Uniqueness and symmetry of equilibria in a chemotaxis model. J. Reine Angew. Math. **654**, 83–124 (2011)
48. Hoshino, D., Koshikawa, N., Suzuki, T., Quaranta, V., Seiki, M., Ichikawa, K.: Establishment of computational model for MT1-MMP dependent ECM degradation and intervention strategies. PLoS Comput. Biol. **8**(4), e1002479 (2012)
49. Hoshino, H., Kawashima, S.: Asymptotic equivalence of reaction-diffusion system to the corresponding system of ordinary differential equations. Math. Model. Meth. Appl. Sci. **5**, 813–834 (1995)
50. Hoshino, H., Kawashima, S.: Exponential decaying component of global solution to a reaction-diffusion system. Math. Model. Appl. Sci. **8**, 897–904 (1998)
51. Hoshino, H., Yamada, Y.: Solvability and smoothing effect for semilinear parabolic equations. Funkcialaj Ekvacioj **34**, 475–494 (1991)
52. Hoshino, H., Yamada, Y.: Asymptotic behavior of global solutions for some reaction-diffusion systems. Nonlinear Anal. **23**, 639–650 (1994)
53. Ichikawa, K., Suzuki, T., Murata, N.: Stochastic simulation of biological reactions, and its applications for studying actin polymerization. Phys. Biol. **7**, 046010 (2010)
54. Ichikawa, K., Rouzimaimaiti, M., Suzuki, T.: Reaction diffusion equation with non-local term arises as a mean field limit of the master equation. Discrete Continuous Dyn. Syst. **S5–1**, 105–126 (2012)
55. Ishihara, S., Otsuji, M., Mochizuki, A.: Transient and steady state of mass-conserved reaction-diffusion systems. Phys. Rev. E **75**, 015203 (2007)
56. Jäger, W.J., Luckhaus, S.: On explosion of solutions to a system of partial differential equations modelling chemotaxis. Trans. Amer. Math. Soc. **329**, 819–824 (1992)
57. Jimbo, S., Morita, Y.: Lyapunov function and spectrum comparison for a reaction-diffusion system with mass conservation. J. Differ. Equ. **255**, 1657–1683 (2013)
58. Joyce, J.A., Pollard, J.W.: Microenvironmental regulation of metastasis. Nat. Rev. Cancer **9**, 239–252 (2009)
59. Karagiannis, E.D., Popel, A.S.: A theoretical model of type I collagen proteolysis by matrix metalloproteinase (MMP)2 and membrante type 1 MMP in the presence of tissue inhibitor of metalloproteinase 2. J. Biol. Chem. **279**(37), 39105–39114 (2004)
60. Karali, G., Suzuki, T., Yamada, Y.: Global-in-time behavior of the solution to a Gierer-Meinhardt system. Discrete Continuous Dyn. Syst. **33**, 885–900 (2013)
61. Kavallaris, N.I., Suzuki, T.: On the finite-time blowup of parabolic equation describing chemotaxis. Differ. Integr. Equ. **20**, 293–308 (2007)
62. Kavallaris, N.I., Suzuki, T.: Non-local reaction-diffusion system involved by reaction radius I. IMA J. Appl. Math. **78**, 614–632 (2013)
63. Kavallaris, N.I., Suzuki, T.: Non-local reaction-diffusion system involving reaction radius II: rate of convergence. IMA. J. Appl. Math. **79**, 161–176 (2014)
64. Kawasaki, S., Minerva, D., Suzuki, T., Itano, K.: Exacting solvable units of variables in nonlinear ODEs of ECM degradation pathway networks, Comp. Math. Mech. Medicine (to appear)
65. Keener, J.P.: Activator and inhibitors in pattern formation. Stud. Appl. Math. **59**, 1–23 (1978)
66. Keller, E.F., Segel, L.A.: Initiation of slime mold aggregation viewed as an instability. J. Theor. Biol. **26**, 399–415 (1970)
67. Kerr, R.A., Bartol, T.M., Kaminsky, B., Dittrich, M., Chang, J.J., Baden, S.B., Sejnowski, T.J., Stiles, J.R.: Fast Monte Carlo simulation methods for biological reaction-diffusion systems in solution and on surfaces. SIAM J. Sci. Comp. **30**, 3126–3149 (2008)

68. Koshikawa, N., Giannelli, G., Cirulli, V., Miyazaki, K., Quaranta, V.: Role of cell surface metalloprotease MT1-MMP in epithelial cell migration over laminin-5. J. Cell Biol. **148**, 615–624 (2000)
69. Koshikawa, N., Minegishi, T., Sharabi, A., Quaranta, V., Seiki, M.: Membrane type matrix metalloproteinase-1 (MT1-MMP) is a processing enzyme for human laminin gamma 2 chain. J. Biol. Chem. **280**(1), 88–93 (2005)
70. Koshikawa, N., Mizushima, H., Minegishi, T., et al.: Membrane type 1-matrix metalloproteinase cleaves off the NH2-terminal portion of heparin-binding epidermal growth factor and converts it into a heparin-independent growth factor. Cancer Res. **70**, 6093–6103 (2010)
71. Kurokiba, M., Ogawa, T.: Finite time blow-up of the solution for a nonlinear parabolic equation of drift-diffusion type. Differ. Integr. Equ. **16**, 427–452 (2003)
72. Kubo, A., Hoshino, H., Suzuki, T.: Asymptotic behavior of solutions to a parabolic ode system. In: Choe, H.J., Lin, C.S., Suzuki, T., Wei, J. (eds.) Proceedings of the 5th East Asia PDE Conference Gakkotosyo, pp. 121–136, Tokyo (2005)
73. Kubo, A., Saito, N., Suzuki, T., Hoshino, H.: Mathematical models of tumour angiogenesis and simulations. Kokyuroku RIMS **1499**, 135–146 (2006)
74. Kubo, A., Suzuki, T.: Asymptotic behavior of the solution to a parabolic ODE system modeling tumor growth. Differ. Integr. Equ. **17**, 721–736 (2004)
75. Kubo, A., Suzuki, T.: Mathematical models of tumour angiogenesis. J. Comput. Appl. Math. **204**, 48–55 (2007)
76. Ladyženskaja, O.A., Solonikov, V.A., Ural'ceva, N.N.: Linear and Quasilinear Equations of Parabolic Type. American Mathematical Society. Providence, R.I. (1968)
77. Laffaldano, G., Caputo, M., Marino, S.: Experimental and theoretical memory diffusion of water in sand. Hydro. Earch Sys. Sci. Discuss. **2**, 1329–1357 (2005)
78. Latos, E., Suzuki, T.: Global dynamics of reaction-diffusion system with mass conservation. J. Math. Anal. Appl. **411**, 107–118 (2014)
79. Latos, E., Suzuki, T., Yamada, Y.: Transient and asymptotic dynamics of a prey-predator system with diffusion. Math. Meth. Appl. Sci. **35**, 1101–1109 (2012)
80. Levine, H.A., Sleeman, B.D.: A system of reaction and diffusion equations arising in the theory of reinforced random walks. SIAM J. Appl. Math. **57**, 683–730 (1997)
81. Lions, J.L.: Quelques Méthodes de Résolution des Problèmes aux Limites Non Linéaires. Dunod Gauthier Villars, Paris (1969)
82. Lions, P.L.: The concentration-compactness principle in the calculus of variation. The locally compact case, Part I. Ann. Inst. H. Poincaré, Analyse Nonlinéaire **1**, 109–145 (1984)
83. Luckhaus, S., Sugiyama, Y., Velázquez, J.J.L.: Measure valued solutions of the $2S$ Keller-Segel system. Arch. Rational Mech. Anal. **206**, 31–80 (2012)
84. Mainardy, F., Gorenflo, R.: On Mittag-Lefler-type functions in fractional evolutional process. J. Comp. Appl. Math. **118**, 283–299 (2000)
85. Masuda, K.: On the global existence and asymptotic behavior of solutions of reaction-diffusion equations. Hokkaido Math. J. **12**, 360–370 (1983)
86. Morita, Y.: Spectrum comparison for a conserved reaction-diffusion system with a variational property. J. Appl. Anal. Comput. **2**, 57–71 (2012)
87. Morita, Y., Ogawa, T.: Stability and bifurcation of non-constant solutions to a reaction-diffusion system with conservation of a mass. Nonlinearity **23**, 1387–1411 (2010)
88. Murray, J.D.: Mathematical Biology, I. An Intorduction, third edn. Springer, New York (2001)
89. Nagai, T.: Blow-up of radially symmetric solutions to a chemotaxis system. Adv. Math. Sci. Appl. **5**, 581–601 (1995)
90. Nagai, T.: Blowup of nonradial solutions to parabolic-elliptic systems modeling chemotaxis in two-dimensional domains. J. Inequal. Appl. **6**, 37–55 (2001)
91. Nagai, T.: Convergence to self-similar solutions for a parabolic-elliptic system of drift-diffusion type in R^2. Adv. Differ. Equ. **16**, 839–866 (2011)
92. Nagai, T., Senba, T., Yoshida, K.: Application of the Trudinger-Moser inequality to a parabolic system of chemotaxis. Funkcialaj Ekvacicj **40**, 411–433 (1997)

93. Nagumo, J., Arimoto, S., Yoshizawa, S.: An active pulse transmission line simulating nerve axon. Proc. I.R.E. **50**, 2061–2070 (1962)
94. Naito, Y., Suzuki, T.: Self-similarity in chemotaxis systems. Colloquium Math. **111**, 11–34 (2008)
95. Nanjundiah, V.: Chemotaxis, signal relayin, and aggregation morphology. J. Theor. Biol. **42**, 63–105 (1973)
96. Nishiura, Y.: Global structure of bifurcating solutions of some reaction-diffusion systems. SIAM J. Math. Anal. **13**, 555–593 (1982)
97. Ohta, T., Kawasaki, K.: Equilibrium morphology of block copolymer melts. Macromolecules **19**, 2621–2632 (1986)
98. Okubo, A., Levin, S.A.: Diffusion and Ecological Problems: Modern Perspectives, 2nd edn. Springer, New York (2001)
99. Olson, M.W., Gervasi, D.C., Mobashery, S., Fridman, R.: Kinetic analysis of the binding of human matrix metalloproteinase-2 and -9 to tissue inhibitor of metalloproteinase (TIMP)-1 and TIMP-2. J. Biol. Chem. **272**, 29975–29983 (1997)
100. Othmer, H.G., Dumber, S.R., Alt, W.: Models of dispersal in biological systems. J. Math. Biol. **26**, 263–298 (1988)
101. Othmer, H.G., Stevens, A.: Aggregation, blowup, and collapse: the ABC's of taxis in reinforced random walk. SIAM J. Appl. Math. **57**, 1044–1081 (1997)
102. Otsuji, M., Ishihara, S., Co, C., Kaibuchi, K., Mochizuki, A., Kuroda, S.: A mass conserved reaction-diffusion system captures properties of cell polarity. PLoS Comput. Biol. **3**, e108 (2007)
103. Pawłow, I., Suzuki, T., Tasaki, S.: Stationary solutions to a strain-gradient type thermoviscoelastic system. Differ. Integr. Equ. **25**, 289–340 (2012)
104. Penrose, O., Fife, P.C.: Thermodynamically consistent models of phase-field type for the kinetics of phase transition. Physica D **43**, 44–62 (1990)
105. Podlubny, I.: Fractional Differential Equations. Acadmic Press, New York (1999)
106. Rascle, M.: Sur une équation intégro-différentielle non linéaire issue de la biologie. J. Differ. Equ. **32**, 420–453 (1979)
107. Riley, M.R., Buettner, H.M., Muzzio, F.J., Reyes, S.C.: Monte Carlo simulation of diffusion and reaction in two-dimensional cell structures. Biophys. J. **68**, 1716–1726 (1995)
108. Rothe, F.: Global Solutions of Reaction-Diffusion Systems. Lecture Notes in Math, vol. 1072. Springer, Heidelberg (1984)
109. Saitou, T., Rouzimaimaiti, M., Koshikawa, K., Seiki, M., Ichikawa, K., Suzuki, T.: Mathematical modling of invadopodia formation. J. Theor. Biol. **298**, 138–146 (2012)
110. Sakurai-Yagata, M., Recchi, C., Le Dez, G., Sibarita, J.B., Daviet, L., Camonis, J., D'Souza-Schorey, C., Chavrier, P.: The interaction of IQGAP1 with the exocyst complex is required for tumor cell invasion downstream of Cdc42 and RhoA. J. Cell Biol. **181**(6), 985–998 (2008)
111. Sato, H., Takino, T., Okada, Y., Cao, J., Shinagawa, A., Yamamoto, E., Seiki, M.: A matrix metalloproteinase expressed on the surface of invasive tumor cells. Nature **370**, 61–65 (1994)
112. Schenk, S., Hintermann, E., Bilban, M. et. al.: Binding to EGF receptor of a laminin-5 EGF-like fragment liberated during MMP-dependent mammary gland involution. J. Cell Biol. **161**, 197–209 (2003)
113. Seki, K., Wojcik, M., Tachiyac, M.: Fractional reaction-diffusion equation. J. Chem. Phys. **119**, 2165–2170 (2003)
114. Senba, T.: Blow-up of radially symmetric solutions to some systems of partial differential equations modelling chemotaxis. Adv. Math. Sci. Appl. **7**, 79–92 (1997)
115. Senba, T.: Type II blowup of solutions to a simplified Keller-Segel system in two dimensions. Nonlinear Anal. **66**, 1817–1839 (2007)
116. Senba, T., Suzuki, T.: Some structures of the solution set for a stationary system of chemotaxis. Adv. Math. Sci. Appl. **10**, 191–224 (2000)
117. Senba, T., Suzuki, T.: Chemotactic collapse in a parabolic-elliptic system of mathematical biology. Adv. Differ. Equ. **6**, 21–50 (2001)

118. Senba, T., Suzuki, T.: Parabolic system of chemotaxis; blowup in a finite and in the infinite time. Meth. Appl. Anal. **8**, 349–367 (2001)
119. Senba, T., Suzuki, T.: Weak solutions to a parabolic-elliptic system of chemotaxis. J. Funct. Anal. **191**, 17–51 (2002)
120. Sire, C., Chavanis, P.-H.: Thermodynamics and collapse of self-gravitating Brownian particles in D dimensions. Phys. Rev. E **66**, 046133 (2002)
121. Sung, J., Barkai, E., Silbey, R.J., Lee, S.: Fractional dynamics approach to diffusion-assisted reactions in disordered media. J. Chem. Phys. **116**, 2338–2341 (2002)
122. Suzuki, T.: Free Energy and Self-Interacting Particles. Birkhäuser, Boston (2005)
123. Suzuki, T.: Mean Field Theories and Dual Variation, 2nd edn. Atlantis Press, Paris (2015)
124. Suzuki, T.: Exclusion of boundary blowup for $2D$ chemotaxis system provided with Dirichlet boundary condition for the Poisson part. J. Math. Pure. Appl. **100**, 347–367 (2013)
125. Suzuki, T.: Blowup in infinite time for $2D$ Smoluchowski-Poisson equation. Differ. Integr. Equ. **28**, 601–630 (2015)
126. Suzuki, T.: Almost collapse mass quantization in $2D$ Smoluchowski-Poisson equation. Math. Meth. Appl. Sci. **38**, 3587–3600 (2015)
127. Suzuki, T.: Residual vanishing for blowup solutions to $2D$ Smoluchowski-Poisson equation. arXiv:1502.01795
128. Suzuki, T., Senba, T.: Applied Analysis: Mathematical Methods in Natural Science, 2nd edn. Imperial College Press, London (2011)
129. Suzuki, T., Takahashi, R.: Global in time solution to a class of tumor growth systems. Adv. Math. Sci. Appl. **19**, 503–524 (2009)
130. Suzuki, T., Tasaki, S.: Stationary Fix-Caginalp equation with non-local term. Nonlinear Anal. **71**, 1329–1349 (2009)
131. Suzuki, T., Tasaki, S.: Stationary solutions to a thermoelastic system on shape memory materials. Nonlinearity **23**, 2623–2656 (2010)
132. Suzuki, T., Yamada, Y.: A Lotka-Volterra system with diffusion. In: Aiki, T., Fukao, T., Kenmochi, N., Niezgódka, M., Ôtani, M. (ed.) Proceedings of the 5th Polish-Japanese days, Gakuto International Series on Mathematical Sciences and Applications, Nonlinear Analyisi in Interdisciplinary Sciences, vol. 36, pp. 215–236 (2013)
133. Suzuki, T., Yamada, Y.: Global-in-time behavior of Lotka-Volterra system with diffusion. Indiana Univ. Math. J. **64**, 181–216 (2015)
134. Toland, J.F.: Duality in nonconvex optimization. J. Math. Anal. Appl. **66**, 399–415 (1978)
135. Toland, J.F.: A duality principle for non-convex optimization and the calculus of variations. Arch. Rational Mech. Anal. **71**, 41–61 (1979)
136. Tonegawa, Y.: On the regularity of chemical reaction interface. Comm. Partial Differ. Equ. **23**, 1181–1207 (1998)
137. Toth, M., Bernardo, M.M., Gervasi, D.C., Soloway, P.D., Wang, Z., Bigg, H.F., Overall, C.M., DeClerck, Y.A., Tschesche, H., Cher, M.L., Brown, S., Mabashery, S., Fridman, R.: Tissue inhibitor of metalloprotenase (TIMP)-2 acts synergistically with synthetic matrix metallo-proteinase (MMP) inhibitors but not with TIMP-4 to enhance the (Membrane type 1)-MMP-dependent activation of pro-MMP-2. J. Biol. Chem. **275**, 41415–41423 (2000)
138. van Zon, J.S., ten Wolde, P.R.: Simulating biochemical networks at the particle level and in time and space: Green's function reaction dynamics. Phys. Rev. Lett. **94**, 128103 (2005)
139. Watanabe, A., Hoshino, D., Koshikawa, N., Seiki, M., Suzuki, T., Ichikawa, K.: Critical role of transient activity of MT1-MMP for ECM degradation in invadopodia. PLoS Comput. Biol. **9**(5), e1103086 (2013)
140. Wei, J., Wintner, M.: Stability of monotone solutions for the shadow Gierer-Meihardt system with finite diffusivity. Differ. Integr. Equ. **16**, 1153–1180 (2003)
141. Wei, J., Wintner, M.: Existence and stability analysis of symmetric patterns for the Gierer Meinhardt system. J. Math. Pures Appl. **83**, 433–476 (2004)
142. Weijer, C.: Dictyostelium morphogenesis. Curr. Opin. Genet. Dev. **14**, 392–398 (2004)
143. Wolansky, G.: On the evolution of self-attracting clusters and applications to semilinear equations with exponential nonlinearity. J. Anal. Math. **59**, 251–272 (1992)

144. Wolansky, G.: On steady distributions of self-attracting clusters under friction and fluctuations. Arch. Rational Meth. Anal. **119**, 355–391 (1992)
145. Wolansky, G.: A concentration theorem for the heat equation. Monatsh. Math. **132**, 255–261 (2001)
146. Wolansky, G.: Multi-components chemotactic system in the absence of conflicts. Euro. J. Appl. Math. **13**, 641–661 (2002)
147. Yamaguchi, H., Pixley, F., Condeelis, J.: Invadopodia and podosomes in tumor invasion. Euro. J. Cell Biology **85**, 213–218 (2006)
148. Yanagida, Y.: Reaction-diffusion systems with skew-gradient structure. Math. Appl. Anal. **8**, 209–226 (2001)
149. Yanagida, Y.: Mini-maximizers for reaction-diffusion systems with skew-gradient structure. J. Differ. Equ. **179**, 311–335 (2002)
150. Yang, Y., Chen, H., Liu, W.: On existence and non-existence of global solutions to a system of reaction-diffusion equations modeling chemotaxis. SIAM J. Math. Anal. **33**, 763–785 (1997)

Index

A

Actin, 13, 21, 22, 24, 59
  actin polymerization, 57
  F-actin, 13, 57, 60
  G-actin, 59–61

B

Biological model
  (A), 35, 36, 92, 99, 100
  (B), 35, 36, 38, 92
  Chaplain-Anderson, 19
  invasion, 19
  Keller-Segel, 16
Brownian motion, 115

C

Cancer
  angiogenesis, 30, 31
  cell deformation, 13, 106
  in-travasation, 106
  invadopodia, 106
  invasion, 1, 13, 19
  membrane degradation, 1
  microenvironment, 106
  pathway, 1
  podosomes, 106
  tumor, 30
Cell biology
  basal membrane, 1
  enzyme, 1
  ligand, 19

plasma membrane, 21–23
protease, 1
receptor, 19
Chemical reaction
  mass action, 3, 5, 8, 17, 20, 52
  mass conservation, 4, 6, 8–10, 14, 19, 24–26, 32–34, 36–39, 87, 91, 94, 96
  mass quantization, 35
  mass reaction, 54, 55
  Michaelis Menten process, 32
  order parameter, 35
Chemotaxis, 16, 31, 34–36, 91, 94, 106, 108, 112, 114

D

Dictyostelium discoideum, 35, 106

E

ECM, 13, 20–22
  ECM degradation, 13, 22
Eigenvalue, 99, 119
Equation
  Allen-Cahn, 22
  Euler-Lagrange, 92, 97
  FitzHugh-Nagumo, 100
  Fix-Caginalp, 96
  Gierer-Meinhardt, 99
  Poisson, 34
  Smoluchowski, 106, 107
  Smoluchowski-Poisson, 33, 36, 37, 65, 100

© Springer Nature Singapore Pte Ltd. 2017
T. Suzuki, *Mathematical Methods for Cancer Evolution*,
Lecture Notes on Mathematical Modelling in the Life Sciences,
DOI 10.1007/978-981-10-3671-2

F
Formula
Einstein, 41, 43, 57, 62
Liouville, 24–26
Free energy
Helmholtz, 35–37
Ohta-Kawasaki, 101

H
Haptotaxis, 31

L
Lagrangian, 92, 98, 100, 101
Law
Newton's third, 37
second law of thermodynamics, 37

M
Mean field limit, 39, 40, 42, 53, 58, 59, 63
MMP, 13, 20–22
MMP2, 1
MT1-MMP, 1
TIMP2, 1
Modeling
bottom-up modeling, 34
top down modeling, 33, 94

P
Phase, 94
phase tansition, 96

S
Self-organization, 35
State
quasi-stationary, 32
Statistical mechanics
entropy, 37
free energy, 35–38
Stochastic process
Poisson, 50, 55
Stochastic simulation, 52
System
Gierer-Meinhardt, 65
Hamilton, 101
Keller-Segel, 32
Lotka-Volterra, 65, 102
ODE, 103
Smoluchowski-ODE, 18, 33

T
Thermodynamical model
FitzHughNagumo, 98
Gierer Meinhardt, 98
Lotka-Volterra, 101

V
Variational duality
Kuhn-Tucker, 98, 101
semi-duality, 97
semi-Toland, 96
Toland, 93, 94
Variational structure
Hamilton, 93, 105
Kuhn-Tucker, 93
Toland, 93

Printed in the United States
By Bookmasters